THE HISTORY OF GEOGRAPHY

TRANSLATIONS OF SOME FRENCH AND GERMAN ESSAYS

Edited
by

GARY S. DUNBAR

Undena Publications
1983

This book is dedicated
to my mentors at Louisiana State University—
John Duffy, William Haag, Fred Kniffen,
Robert West, and the late Richard Russell—
titans all!

Library of Congress Card Number: 83-51533
ISBN: 0-89003-148-7/paper

Undena Publications, P.O. Box 97, Malibu, CA 90265, U.S.A.

TABLE OF CONTENTS

EDITOR'S INTRODUCTION

The history of geography, as an academic subdiscipline conforming to modern canons of scholarship, is still in its infancy. Not even the most optimistic observer could say that it has proceeded beyond adolescence. We are still in the process of assembling building blocks from which a substantial edifice might in time be constructed. Only in the last few years has an "invisible college" of specialists in the history of geography emerged, and it will be some time before this college has produced a large and comprehensive body of literature. In the meantime, we must gather the existing materials in coherent order so as to point the way to future research.

There are surprisingly few essays of a general nature in the history of geography, and they are not so common in English as in French and German. This collection of translations was prepared for the use, in the first instance, of the postgraduate students in my course in the History of Geography in 1979. I felt that such a compendium would be of great practical value in helping novitiates chart their way. I hope that these translations, despite whatever infelicities may remain, will accurately convey the meanings that the authors intended. I shall be delighted to have readers point out errors that my ignorance has allowed to persist. The translations are mostly literal, rather than free, so that a reader can quite easily trace a word or phrase of special interest back to the original text. I have made a few minor corrections but no substantial revisions. In a few places, I have altered the original punctuation or word order, but only to make the English less cumbersome.

I prepared the French translations myself, and the initial rough translations of the German selections were done by Mrs. Helge Prosak (Beck) and Mrs. Barbara Gross (Wagner and Hettner). I hope that in trying to smooth out the German translations and to correct some geographical terms and concepts that were unfamiliar to the translators I have not induced greater error. As President Carter could attest after his return from Poland, translation is a difficult art. For me, no translation is easy, whether it be from French or German or even from British English into American English. Early in my career, I even encountered difficulties in translating Virginia English into my native Upstate New York dialect.

Anthologies are peculiar constructions, so that compilers should be prepared to explain or defend their choices of materials. I chose these six papers primarily because I wanted essays that were "factual" treatments of the history of geography and not strongly theoretical or methodological. Taken altogether, I think that they give the reader a good overview of the field without pulling him into the quagmires of philosophy or methodology. Three of the essays were written by eminent geographers of the earlier part of the 20th century (Wagner, Hettner, and de Martonne), and the others represent the work of active researchers in the history of geography (Beck, Pinchemel, and Claval). I chose Beck's 1957 paper instead of one of his more recent essays because of its factual detail, but readers would be well advised to go on to other publications of this prolific author. The selection of French and German papers does not mean that there are no important materials in English and other languages. It is simply that, apart from English, French and German are the languages of greatest general utility in the study of the history of geography. The majority of our postgraduate students do not know enough French or German to be able to appreciate the wealth of materials in those languages. Some of them gain such expertise during the course of their postgraduate education, but they are normally required to take a course in the History of Geography at the very beginning of their careers. Even those students who are already fluent in French and German, or those who have achieved at least the low level of proficiency that I myself possess, will still find value in this compilation, especially in the bibliographical and biographical notes. I am primarily concerned with the modern period in geography, which began with the two German "fathers" or founders, Alexander von Humboldt (1769-1859) and Carl Ritter (1779-1859), so that I have not included biographical notes on earlier geographers.

The living authors (Professors Claval, Pinchemel, and Beck) and their publishers have granted permission to publish these translations. The authors have read my translations and approved them generally, but they cannot be blamed for any of the barbarisms in the English text. The manuscript was typed at the UCLA Word Processing Center, and Mr. Noel Diaz, Principal Illustrator in the Department of Geography, prepared the title page and chapter headings.

Gary S. Dunbar
University of California
Los Angeles
November 1983

THE EVOLUTION OF GEOGRAPHY

Emmanuel de Martonne

(This is a translation of Chapter 1,
"Evolution de la géographie," of Part One of
de Martonne's Traité de géographie physique,
4th edition, vol. 1 [Paris: Librairie Armand
Colin], 1925, pp. 3-26. This chapter was
essentially unchanged from the first edition
of 1909. I have omitted the two maps that
accompanied de Martonne's essay.)

What is geography? That is the first question that seems to be asked
at the beginning of lessons in general geography. To try to answer it with
some precision, it is necessary to see how that science was formed and how
it became by a slow evolution one of the most complex that the human mind
had invented. To wish to define a science by principles posed a priori and
to assign exact limits to its field of investigation is always a risky
enterprise. It seems that the more human knowledge progresses, the more
appear the connections that link the various sciences, like branches
issuing from a common trunk. The circumstances that determine the
attribution of a certain order of researches to the practitioners of a
certain science are almost always fortuitous. The history of geography
offers more than one example of that. Any a priori definition that does
not take account of the natural evolution of things risks remaining without
influence, or risks exercising a bad influence.

To render an account of the stages that geographical science has
passed through, of the vicissitudes that its development has suffered, and
of the transformations—slow, but continuous—that have led it to the high
degree of organization where we see it today, there is, it seems, the
surest and most prudent method of recognizing the intimate principle of it
and of elucidating what it is and what it ought to be.

I.--The Geography of the Ancients

It is often said that geography is a new science. That is true, if
one wishes to speak of the ensemble of scientific researches that have been
grouped under that name for the last fifty years. But geographical science
taken in its largest sense, as science of the earth, is one of the most
ancient branches of human knowledge. It responds to one of the most
essential needs of human nature, as soon as man was raised above the savage
life: that of fixing the memory of the places that surround us, within a
radius whose length varies with our needs and our means of locomotion.

Didn't the Polynesians have some kinds of rudimentary geographical maps in the form of curved sticks? (Droeber, 1)

If the name of geography, geographia, seems to have been invented by the Alexandrians, the thing itself appeared with the first Greek writers, from the time that thought awakened and the circle of experience enlarged beyond the horizon of the hamlet or the town. It is even curious to note that one can recognize from the beginning, not only the object and the principal directions of geography, but even the difficulties and the essential conditions of development of that science.

The Alexandrians made Homer the first geographer. It is certain that the homeric poems were appreciated for the exactness of their depiction of maritime life and the geographical conditions of distant regions, no less than for their literary beauty. But the first truly conscious geographer was Herodotus, the historian who enlarged the field of local chronicles, and whom the study of a great war brought in touch with lands as distant as they were different. His travels in Egypt, Thrace as far as the Hellespont, Phoenicia, and up to Babylonia are known. He represented the descriptive trend of geography--what we call regional geography.

At the same time, and even a little before, there was revealed, in the Ionian cities where Greek thought first arose, another side of geography, that which looks at the earth as a whole and bears the name of general geography. The Ionian natural philosophers, of whom Thales of Miletus was the most glittering name, began as early as the 6th century B.C. to get involved with the problems of terrestrial physics: form, dimension, and position of the earth in space. Thales, inheritor of the astronomical lore of the Egyptians and Babylonians, announced the sphericity of the earth, which our Middle Ages felt compelled to doubt.

Thus, from the beginning appeared the two essential points of view of geography considered as the science of the earth: general geography and regional geography are found throughout Antiquity contending for the attention of scholars. When one thinks of the weakness of the means of investigation that were at the disposal of the ancients, one is astonished that general geography was able to make such progress. Born with the Ionians, summarized rather than developed by Aristotle, and considerably enlarged by the Alexandrians, it grappled with the loftiest problems of geophysics. Not only did it prove the sphericity of the earth, but it measured the dimensions of it with Eratosthenes (c. 230 B.C.). It looked at hydrological and climatological problems: the continuity of the oceans, the theory of climatic zones, and the origin of rivers and their floods, principally the floods of the Nile. All that was unfortunately drawn from considerations too remote from experience; truth was mixed with error, in such proportions that it was then impossible to unravel them. Also one sees from time to time spirits with a practical tendency stirring and, leaving all theories aside, turning toward the exact descriptions of known lands.

That reaction, which brought regional geography back into favor, was particularly marked among the Roman geographers or inspired by Roman ideas. Polybius gave the signal for it. Strabo, whose work we have preserved almost in its entirety, was the writer who showed what this trend could yield. One observes that regional and descriptive geography is much more human, more attentive to ethnography, to the migrations of peoples, and to customs and institutions, while general geography is more physical, more exact, or at least more concerned with mathematical precision. Ptolemy (2nd century), who represents a reaction against the descriptive trend, would bring to bear all his effort on the exposition of general notions of mathematical geography and on the fixing of the geographical positions of a certain number of points.

It was important to mark well that double face of geographical science, which is seen from the beginning. For a long time the dualism would persist. One would see these two branches of the same science developing side by side without touching or penetrating each other. Modern geography would exist from the day when the rapprochement and fusion would be brought about by powerful intellects, such as Humboldt and Ritter.

A second point still merits our attention. The development of geography, hindered by this dualism, is still held back by the fact that it is subordinated to outside circumstances. From antiquity, in effect, one can observe that the progress of geography was tied on the one hand to that of the knowledge of the globe, that is to say, to wars or political events, and on the other hand to the progress of the sciences.

Every important enlargement of the known world gave new scope to descriptive geography. The first example was given by the conquest of Alexander, who suddenly extended the geographical horizon to India. The Roman conquests had the same result.

Geography appeared also, from antiquity, as tributary to certain sciences. It was to the progress of astronomy that the fate of general mathematical geography was tied, and it was by astronomical considerations that the Ionians proved the sphericity of the earth and that Eratosthenes was led to measure the dimensions of it. Likewise regional or descriptive geography was born, with Herodotus, out of the requirements of history.

The dualism of geographical thought and its dependence on social and political evolution as well as on the progress of various sciences are the factors that for a long time would govern the development of geography.

II.--The Middle Ages and the Renaissance

The Middle Ages were for geography, as for the majority of the sciences, a period of eclipse or even of recession. The period is marked especially by a complete decline of general geography. It was by the Arabs that the hearth was maintained and geographical activity was still manifested; but, whatever the importance of the Arab geographers was, it must be recognized that they were almost completely turned toward descriptive geography. All were great travelers: Massudi of Baghdad, the author of Prairies d'or, who died in 957, had traveled through Palestine,

Persia, Armenia, Syria, Egypt, North Africa, and Spain; Mohammed el Edrisi, born in the 12th century in Morocco, had visited the coasts of France and England and the interior of Morocco and Asia; Ibn Batuta, also Moroccan, traveled in the 14th century through northern and western Africa as far as the Niger, western Asia, India, China, and southern Russia. In the descriptions of these indefatigable travelers, human geography takes the principal place, with many historical and political details. The notions of general geography, which often figured in the title of their books, were drawn from Ptolemy, sometimes badly understood (Lelewel, 13).

The Renaissance marked for geography, as for nearly all the branches of human knowledge, an era of renewal and of feverish activity. It was the age of the great voyages, which revealed unknown worlds, and of great scientific discoveries, which gave new foundations to all learning. It was a period of crisis, from which geography could have emerged full formed. In reality three great occurrences marked this unique moment: 1. a prodigious enlargement of the geographical horizon; 2. the great development of cartography; and 3. the progress of the physical sciences auxiliary to geography.

Before the century that was justly named the age of the great geographical discoveries, some distant travels had already remarkably extended the knowledge of the globe. We have spoken of the peregrinations of the Arab geographers. In the 13th century Marco Polo had traveled throughout nearly all of Asia, sojourned ten years at the court of the Mongol khans, and visited Indochina, India, and the Spice Islands. But, a complete stranger to astronomical methods, he did not report any determination, even approximate, which could serve to fix these new lands on the map. Whence came the error of longitude that is reflected in the globe of Martin Behaim (1492), reducing to 120° the distance between [western] Europe and [eastern] Asia, an error shared by Toscanelli, who inspired Christopher Columbus. The great voyagers of the 16th century were, on the contrary, sailors who were used to observations and to nautical precision.

The succession of discoveries in the thirty years that stretched from 1492 to 1522 was something spectacular. First there are the four voyages of Christopher Columbus (1492-1493-1498-1502), who revealed the New World, while believing that he had reached eastern Asia. At nearly the same time, Vasco da Gama, following a route already frequented by the Portuguese, all along the west coast of Africa, accomplished the voyage around the continent, which had been begun by Dias, and reached the shores of India (1497-1499). Vespucci and Vincent Pinzon reconnoitred the northern coast of South America. Cortes penetrated to the heart of Mexico (1519-1521), and Cabot touched Labrador and Newfoundland (1497). Finally, the expedition of Magellan accomplished from 1519 to 1522 the first circumnavigation of the world, in rounding the extreme tip of South America. In thirty years the geographical horizon, which had not gone beyond 60° in latitude by 100° in longitude, was extended to embrace nearly all the earth.

One can easily imagine what a revolution could operate in men's minds with such a surprising succession of discoveries. It is here that one can best grasp the connections that unite the history of geography to that of travel. Old theories, still held by the prejudiced, definitely fall into the dust. Such a belief was that of an uninhabitable torrid zone, which arose out of a priori climatic considerations, already opposed in ancient times by Polybius and Eratosthenes, but still accepted in the Middle Ages, which repudiated the existence of the antipodes in relying on the Bible and the Church Fathers (Kretschmer, 11). But, at the same time, one sees these rapid discoveries introducing new errors. The explorers' narratives were full of the marvelous; legends were being formed (such as Eldorado).

The development of cartography, which signaled the Renaissance, was due in great part to the geographical discoveries. But it is also connected with a material fact that one must not forget: the invention of printing. It is thanks to this new process that one can see multiply the editions of Ptolemy, with the maps being constantly improved. The Middle Ages seem to have known only rather gross representations of the terrestrial globe. The tables engraved on silver or the precious globes were objects of art and luxury whose possession was reserved to kings. The painted mappamundi of Fra Mauro (15th century) ignored parallels and meridians. With printing, the geographical maps, whether established according to Ptolemy or intended to fix the new discoveries, passed into the hands of everyone, and the cartographers were induced to envisage the problem of projections in all its forms.

One knows that the first Latin translation of Ptolemy was that of Jacobo Angelo (Vicenza, 1471) and that the first edition accompanied by maps engraved on copper appeared in Rome in 1478. In Italy, France, and Germany there were trained cartographers conversant with all the discoveries (Gallois, 18). Then the position of primacy passed to Holland and Belgium with Gerhard Kremer, known under the name of Mercator, who redrew the maps of Ptolemy (1578) and rediscovered or invented the principal system of projections. A little later Ortelius inaugurated the great collections of modern maps with his Theatrum Orbis Terrarum, which began to appear in 1570.

Despite the efforts of these savants to reconcile Ptolemaic geography with modern geography, the split still existed. For several centuries already the application of the compass to navigation had permitted the sailors to construct route maps known under the name of Portolans, which, without bothering with latitudes and longitudes, gave much more exact shapes of the Mediterranean coasts than the maps of Ptolemy, which had been spoiled from the outset by an error in the position of Gibraltar. If henceforth the seamen recognized the utility of determination of longitudes and latitudes, while the scholars no longer accepted blindly everything that came from antiquity, it nevertheless happened that two different currents of geographical researches were formed: on the one hand the geography called ancient or historical, on the other modern geography, based on the recent discoveries. Ancient geography was considered as more serious and more scientific. Ortelius turned to it with a marked predilection, and the last years of his life were consecrated to the publication of the Parergon, which he considered as his capital work. One could see that state of mind persisting for a long time; traces of it were found in France until the 19th century.

If one seeks to determine exactly what geography had gained in the period of the Renaissance, one must recognize that its progress was less marked than that of the neighboring sciences on which she depended. The renovation of the knowledge of the earth is due in great part to the progress of astronomy. Without the tables of declination constructed by the Jewish scholars who were the heirs of the Arabs, and permitting the determination of latitude by the observation of the sun above the horizon, would Columbus have dared to set forth in the crossing of the ocean? The application of the telescope to astronomy by Galileo gave the probability of fixing longitudes with more precision. Copernicus, in finding the true position of the earth in the solar system (1543), thereby gave even the first foundations of climatology, based on the most general cosmic facts; it was he who dealt the final blow to the old theory of the two spheres, terrestrial and liquid, on which the whole Middle Ages had subsisted. The progress of nautical science outstripped that of geography. On the map of Holland edited by Mercator in 1585, one finds marine depths indicated for the first time. In their daring voyages, the discoverers of America observed marine currents. The tides attracted attention, and soon their causes would be recognized.

Geography, as we understand it today, still had not been born. However, there appears, with the Cosmography of Sebastian Münster, the first model of those great detailed descriptions of the earth that followed at closer intervals up to the 19th century, being renovated with each new extension of the geographical horizon. This considerable work went through 44 editions from 1544 to 1650, tangible proof of the interest that was awakening in the public for geography. In sum, descriptive regional geography seems especially to have gained in the great movement of the Renaissance. But it was and would yet long remain in its infancy because of its lack of contact with general geography.

The latter, seemingly neglected, was being prepared in relative obscurity through the progress of the physical sciences. In the 17th century it was, for the first time since antiquity, collected together anew in a body of doctrine in the admirable Geographia Generalis of Varenius. Dead at the age of 28, there is no way of knowing how far this powerful mind might have gone. Perhaps he might also have renovated descriptive geography; he had presented a plan for a work of this sort. The guiding principles of his general geography were already elevated well above everything that the ancients had conceived; they are the same that would inspire Humboldt two centuries later. All the physical phenomena found a place in his work, which was conceived on a plan of breadth unknown up to that time. The great divisions of general geography were there clearly marked: oceanography, climatology, and orography. On a large number of facts, he had remarks and ideas that were surprising by their soundness and that often surpassed, by a real prescience, the latest progress of the sciences.

Unfortunately, Varenius's book remained without great influence. The scientific rigor--more apparent than real--of his method and his method of exposition by theorems outraged too many old habits and doubtless prevented the book's spread. It was reprinted several times, notably in 1672 through the care of Newton, translated into French and English; but what was that beside the 44 editions of the Cosmographia of Sebastian Münster? Conceived at first in a truly scientific spirit, this sort of summation of descriptive geography was little by little inflated, with each new edition, with narratives more or less marvelous reported by the explorers, for which the public showed itself avid, to become finally a confused mixture of fantasies and truths. Descriptive geography deviated once again from its path.

III.--Modern Times Up to Humboldt and Ritter

Thus the work of the renovation of geography, of which the Renaissance seemed to be the cause, ran aground. There was still lacking to it the support of the natural sciences, which would be scarcely developed by the 18th century. Varenius's geography was still all physical; thereby it had less hold on minds that were alien to the sciences.

The 17th and 18th centuries saw great discoveries accomplished, but the interior of the continents still remained mysterious, peopled with legends. Considerable progress was made by the physical sciences: Halley drew the first map of winds and sketched the theory of the trade winds (1686); Snellius applied to the surveys a procedure that was even then that of triangulation (1615); the barometer, invented by Torricelli, was useful to Pascal for the famous experiment of Puy-de-Dôme, which contained the principle of the measurement of altitudes. No geographer sought to coordinate these data in the spirit of Varenius.

What was produced was rather significant: the edifice built with a sure hand by that precursor was, in the absence of geographers, invaded by the practitioners of neighboring disciplines. The principal part of the share went to a new science, born in the 18th century and given a name nearly identical to geography--geology. The first geologists, Buffon, Hutton, Deluc, and Leopold von Buch, were occupied almost exclusively with phenomena that are now considered as in the domain of physical geography. This movement has for the history of geography a great importance; it explains that, even now, the lines are so narrow between geology and geography that one has difficulty distinguishing them from each other.

The only branches of geography that were developed in the 17th and 18th centuries were ancient or historical geography and mathematical geography. Historical geography appeared to be tied more and more intimately to cartography. It included illustrious scholars such as Guillaume Delisle and d'Anville, whose great merit was in having brought the critical spirit of the historian into the interpretation of documents of very unequal quality that one then possessed on extra-European countries. Delisle cleansed the map of Africa of all the extravagances that the fantasy of the cosmographers had accumulated there and definitely broke with the error of longitude on the Mediterranean that had been perpetuated since Ptolemy (Sandler, 20).

At the same time, mathematical geography took a new development. The great attempts at the measurement of the globe tried by the Greeks with rudimentary means and that long remained the object of an admiration mixed with amazement, were finally taken up again with perfected apparatus that enabled the progress of the natural sciences to be organized. To France went the honor of having inaugurated that new series of researches with the measurement of the arc of the meridian between Paris and Amiens by Jean Picard (1667-1670) and the missions of Maupertuis and La Condamine, charged with effecting the same operation, the latter in Peru and the former in Lappland (1735-1739). The dimensions and the form of the globe were finally fixed; the flattening at the poles and the bulging at the Equator were recognized. In the same period the engineer Cassini began the colossal work of the first topographic map of France at the scale of 1:86,400. The other European countries followed our example.

In the midst of this historical and mathematical activity, geography as we understand it today was reduced to a very small matter. It suffered and would suffer for a long time still from the fact that its development was limited to what there was in it of the less living and the more abstract. All would change, from the day where a Humboldt would show that geography is above all the science of physical and organic life on the surface of the globe. Up to that time, geography was the work of erudites; it could not attract to itself the attention of the public. Only descriptive geography succeeded in enlarging the circle of its audience; but by what means? In tickling the curiosity by anecdotes or extraordinary tales, in mixing in historical and political details completely foreign to geography, and in multiplying the dry enumerations, on the pretext of being useful to travelers, merchants, and statesmen. Kant's Geography, the only book during the 18th century that came up to the ideas of Varenius, sacrificed itself to these tendencies, in devoting a chapter to natural curiosities (Merkwürdigkeiten). The best essays in descriptive geography were always spoiled with the preoccupation with being useful. These tendencies would remain for a long time. In any case, they were reigning when Humboldt and Ritter appeared (Wisotzky, 21).

The revolution accomplished by these two geniuses would be profound and definitive. By their character and their different qualities, they complemented each other very fortunately. Humboldt (1769-1859) was, by taste and education, a man of science--especially a naturalist--and a great traveler. He traversed, as an attentive observer of all physical and biological facts, a great part of Europe and all of Mexico, Central America, Colombia, and Venezuela. He even went off to Russian Asia in the steps of Pallas. His numerous publications were all of a purely scientific nature, completely exempt from pedagogical preoccupations. Ritter (1779-1859) was an armchair scholar, an historian and philosopher by education. He traveled much less than Humboldt, and he did it to enlarge his ideas rather than to devote himself to observations and study on the spot. He was a professor; his publications came out of his teaching at the University of Berlin, and with him the pedagogical preoccupations were always evident.   Humboldt's merits could not be esteemed too highly. He founded the methods of observation of nearly all the branches of physical geography. He popularized the use of the barometer to determine altitudes and the use of cross-sections and calculations of average elevation to characterize relief. He drew the first map of isotherms and showed the contrast between east coasts and west coasts. Again, it was he who should be considered as the creator of botanical geography based on the physiognomy of plants and their connections with soil and climate. No traveler approached him as an observer. From his five years stay in Central and South America, he brought back materials for publications that followed one after another for twenty years. But Humboldt was not only a naturalist and a traveler, he was a geographer with a breadth of views that has rarely been found again since. To him is due the merit of having first defined and applied the two essential principles that make geography an original science, something other than a composite of the physical and biological sciences. Whatever the phenomenon that he studied--relief, temperature, or plant life--Humboldt was not content with looking at it just for itself, and treating it as a geologist, meteorologist, or botanist would. His philosophical mind went further. He proceeded immediately to the other phenomena that the milieu in which he found himself presented to his view; he went back to the most remote causes and consequences, including even political and historical facts. Nothing showed in a more precise fashion how man depends on soil, climate, and vegetation, how vegetation is a function of physical phenomena, and how the latter themselves depend on each other.

To this first principle, which could be called the principle of causality, Humboldt added another, which might be named the principle of general geography. As he fixed his attention on a geological, biological, or human problem, this great mind did not remain absorbed in the contemplation of the local occurrence; he shifted his eyes toward the other regions where analogous occurrences were observed, and it was always a general law, valid for all similar circumstances, that he sought to extract. The study of no single point seemed to him independent of the knowledge of the whole of the earth. The application of this principle was the definite overturning of the barrier that separated regional geography from general geography and was the reconciliation of these two branches of the same science and their interfertilization. From the day that the significance of this was understood, modern geography was born.

Humboldt's work was unfortunately out of proportion to the influence that he exerted, at least on geography. His works, of a purely scientific nature, were destined for scholars; for rather a long time he remained ignored by the general public. It was in the world of naturalists that he found his disciples, especially in France, where he sojourned for a long time and published part of his works. On the contrary, Ritter's influence on geographers was evident in Germany from the first half of the 19th century, and, even in France, it was more perceptible than that of Humboldt. Ritter's great merit was in having sensed and clearly formulated the principles that Humboldt had applied, rather than expressed dogmatically. The principle of general geography was that which he best illuminated. World position (Weltsstellung) was for him the premier fact to consider in the study of any country. He did not appreciate the principle of causality any less profoundly, but it is necessary to acknowledge that he did not always know how to apply it. Too often his physical geography remained descriptive and without connection with the social and historical facts that he insisted on. There was more than one reason for that. At first Ritter's colossal work, comprising the description of continents as little known as Africa and Asia then were, collided with many uncertainties and obscurities. Moreover, Ritter's scientific education was not on a level with his conceptions; he was neither a naturalist nor a physical scientist, like Humboldt. It was a lesson that the young geographers could not ponder too much. Finally, it must not be forgotten that Ritter remained a philosopher in his soul. The man who had the greatest influence on him was Herder. What was called the teleological idea dominated all his work: the earth was for him the theater of human activity, and man played there the same role as the soul in the body.

Ritter's influence was very great, especially in Germany. It is still notable, and the present movement in the delimitation of geography is inspired by his basic ideas.

IV.--Modern Geography after Humboldt and Ritter

After Humboldt and Ritter, one would have been able to believe that modern geography was definitely founded. But the new ideas sown by these two great minds still had not found a sufficiently well-prepared soil. Humboldt had direct influence only on naturalists; but they saw in his works only the invention of methods of observation and continued to go deeply into each class of researches, without taking heed of the ties that could link them together. The development of the natural and physical sciences exceeded all that could be predicted; but geologists, botanists, physicists, and meteorologists specialized more and more. Ritter was followed more by the geographers, but the disciples took from the master more of his faults than of his good qualities. One repeated the great principles of method, but one knew less and less how to apply them. Regional geography once more slid down the slope of dry descriptions mixed with anecdotes and historical facts (Wisotzky, 21).

However, ancient geography and mathematical geography continued their brilliant development of the 18th century, particularly in France. D'Anville had successors, such as Avezac. Tissot and Germain completely renovated the modern science of projections. This brilliance of historical and mathematical geography in France would be slightly harmful to the introduction of modern geography, which would develop much faster in other countries, such as Germany.

The definitive formation of geography could no longer be delayed. The germination of the seeds deposited by Humboldt and Ritter was accomplished, as soon as favorable circumstances presented themselves, that is to say in the last third of the 19th century. It is not without interest to analyze these circumstances, to see how the scientific atmosphere necessary to geography was created.

First it is necessary to note a considerable enlargement of our knowledge of the earth, in connection with the improvements in means of transport and the colonial thrust of all the European states. If the 19th century did not see discoveries as startling as those of the Renaissance, it had, by a series of efforts that were continuous and not without dangers, transformed the geographical image of the world more than any other epoch. Its work was particularly important in the interior of continents. Outside of the coastal areas, what was known of mysterious Africa before the explorations of Barth and Nachtigal in the Sudan and Sahara, of Speke, Schweinfurth, Stanley, and Emin Pasha in the lands of the Upper Nile and the Congo, and of Livingstone in the Zambezi area? The great mountain chains of Central Asia and their structural relations were still enveloped in mist before the illuminating reconnaissances of Przheval'skii, Obruchev, and Sven Hedin. The Rocky Mountains, with their strange world of enclosed basins, fantastic gorges, and desolate plateaus, were nearly completely unknown before the explorations of Hayden, Powell, and other precursors of the United States Geological Survey. It has justly been remarked that through the novelty of the results and the unexpectedness of the physical phenomena discovered, this penetration of the continents contributed more than any other circumstance to arouse ideas and hasten the blossoming of modern scientific geography (Vidal de la Blache, 26).

One must not, however, neglect the valuable acquisitions due to the great oceanographic explorations. Before the 19th century, we knew almost nothing of the oceans, which nevertheless cover two-thirds of the surface of the globe. The first sounding apparatus permitting the exact measurement of the great depths was constructed by Brooke in 1854. Since 1860, a series of scientific cruises, of which the most celebrated was that of the Challenger, revealed to us the conditions of submarine topography and the gross traits of the distribution of temperatures in the seas. The analysis of movements of ocean masses became possible, and their influence on the climate could be made precise.

It is curious to note that the principal effort of continental and oceanic explorations coincided with the publication of bold and captivating geographical works. It was in 1870 that the geological explorations west of the 100th meridian commenced in the United States. At nearly the same time, Przheval'skii began a series of travels in Central Asia, and the Challenger was launched on a great cruise to traverse all the seas of the world (1873). Between 1870 and 1880 an astonishing series of African explorations occurred: Nachtigal, Schweinfurth, Stanley, Cameron, Serpa Pinto, etc. At the same time appeared, in France, La Terre of Elisée Reclus (1868-1869), and in Germany, Neue Probleme der vergleichenden Erdkunde of Oskar Peschel (1870). By the skilfulness of exposition and the attractiveness of their form, these two books have strongly contributed, in spite of their defects, to expand interest in physical geography in the educated public. The success of Reclus's work, undoubtedly due in great part to the literary form and to the poetic turn of its descriptions, had an effect on the conception of his monumental Géographie Universelle, of which the nineteen volumes succeeded each other with an amazing regularity; and this publication, of uneven scientific worth, must be considered as one of the most powerful instruments of the diffusion of geographical knowledge. The <u>Neue Probleme</u> of Peschel had similar qualities and defects. The physical questions pointed out by Ritter had never yet been the object of so profound and so captivating an examination. Peschel and Reclus must be considered as awakeners of ideas. They had on the public the influence of an excellent professor on his charmed and surprised students. Through them geography was animated and appeared as the science of life at the surface of the earth. It ceased to be a science of abstraction and erudition, and that occurred at the time when thrilling voyages drew the attention of all and revealed new facts in the interior--until then mysterious--of the continents.

In spite of all these favorable circumstances, modern geography doubtless could not have been definitively constituted, if the development of the sciences and the increasingly greater part taken in this development by the civilized nations had not taken up tools of inestimable value that had previously been lacking to them. With the first years of the 19th century began the execution and publication of French and English topographic maps, surveyed by the services of the state. The French military map at 1:80,000 and the English map at 1:63,360 were both finished around 1870. This example was followed by all the European countries, the United States, and the Indian Empire. Before the execution of such works, it could be said that in reality nothing precise was known about landforms. With detailed geological maps, published by the services of the state organized now in all the civilized countries, the interpretation of these forms became a relatively easy task. Knowledge of the atmospheric world could be made precise since isolated meteorological observations were replaced by observations that were regular, coordinated, published, and explicated by the meteorological services of the principal nations. Will we forget how the organization of the official statistical services has permitted human geography to emerge from a vague and uncertain state? If one thinks of the situation of Ritter's successors faced with the study of a European country when there existed neither topographic maps equal to those that we possess nor detailed geological maps accompanied by suggestive commentaries, nor atlases and meteorological compendia, one will no longer be surprised that the development of scientific geography had been somewhat retarded, and one will appreciate the importance of these acquisitions for the development of modern geography.

This development was still favored by circumstances that are not completely independent of the facts already cited. The interest awakened in the general public by exploration had the most fortunate influence on the prosperity of geographical societies. It was undoubtedly not extraneous to the organization of geographical teaching in the universities, which tended more and more to focus all the scientific life of modern countries. This last point is particularly important. As ripe as geographical science was, it only commenced to bear fruit from the day that it took root in university soil, in intimate contact with the sciences in the development of which it ought to be associated. What rather proves it is the considerable advance made in Germany, where university geographical teaching had been organized sooner than in other countries.

V.--The Definition of Geography

Geography can be considered as a full-grown science. The plan sketched by Varenius and developed by Humboldt and Ritter seems nearly realized. The geographical domain, once dismembered for the benefit of the natural sciences, has been put back together again. Perhaps in the ardor of this reconstruction one has gone a little too far: geography has taken on the aspect of an encroaching science with encyclopedic tendencies. Similar occurrences are not rare in the history of human knowledge: philology has passed through that period, and sociology is still in it. Geography is starting to leave it; the preoccupation with limiting the field of geographical studies has come out in more than one author. There is in that a practical interest evident, that of the economy of forces and of the best division of scientific work.

The ties of geography with geology are closest, and it is from this side that one has felt most keenly the need to trace a limit. Richthofen (22) excluded from physical geography the study of the bedrock, reserved to geology. Mackinder tells us: geography is the science of the present explained by the past, geology is the science of the past explained by the present. This last definition has the advantage of showing very well the reciprocal services that these two sciences can and must render. No more than geography can dispense with examining the history of the earth in the light of geology in order to explain the present landforms, geology cannot explain the phenomena of earlier periods in earth history without studying the similar phenomena that we see being produced under our eyes on the land and in the oceans. Thus we see between geology and geography a difference of method rather than of object.

In reality, the exact delimitation of the field of geographical investigations is a chimerical enterprise. This science touches too many sciences, and it has--its history proves it--too much interest in remaining in contact with them for anyone ever to desire that limitation.

The essential thing is to isolate the methodological principles that now seem to be established. The principle of extension has been particularly well brought out by Ratzel (4). A few examples suffice to show its significance: the botanist studies the organs of a plant, its conditions of life, and its position in the classification; if he seeks to determine its area of extension, he says that he is doing botanical geography. The geologist analyzes the mechanism of a volcanic phenomenon for its own sake; he is conscious of doing physical geography when he seeks to determine the distribution of volcanoes. The statistician combines numbers with a view to establishing the course of various demographic phenomena; if he tries to account for the distribution of population, he knows that he is doing human geography.

This geographical point of view is often shown to be fruitful, and no science has to be sorry for adopting it. De Candolle was one of the first to show how even the form of the geographical area of a plant and the direction that its border follows indicate its requirements and its conditions of life. A good map of volcanoes is indispensable in order to study the causes of volcanism. The movements of population that especially interest the statistician could not be studied in a fruitful fashion if one had not succeeded in locating exactly the regions of depopulation and those of growth of population. From the importance of the principle of extension results that of cartography. Without going as far as pretending that geography and cartography are synonymous, one must remark that all study receives a geographical cachet when one seeks to express its results cartographically.

The principle of general geography has been expressed with force by Ritter, and more recently by P. Vidal de la Blache. One can formulate it thus: the geographical study of a phenomenon assumes the constant preoccupation with similar phenomena that appear in other parts of the earth. For example, analysis of the characteristics of the Breton coasts takes on geographical value if we can compare them with similar coasts, so as to show how their particular characteristics are explained by the general principles of the evolution of littoral forms.

The application of this principle assumes the knowledge of the greatest part of the earth; one should not be surprised that one had to wait until the 19th century to gather the first fruits of it. They are the comparisons between the continental similarities that aroused Peschel's attention to problems of physical geography. No one realized better than Humboldt the fruitful union of regional geography and general geography that up to his time always remained separated.

The third principle of geographical method is the principle of causality: never to be content with the examination of a phenomenon without trying to go back to the causes that determine its extent and without investigating its consequences, is to be placed on ground that is not properly that of any of the physical, natural, or social sciences with which geography is in contact. The application of this method was the principal originality of Humboldt's writings. It is in its application that the greatest progress was realized in the 19th century. One has seen it successively vivify each of the branches of geographical science, particularly the study of landforms and anthropogeography.

The description of landforms seems to us inseparable from their explanation, and this explanation implies the study of their past. To the American geographers is owed the merit of having generalized this historical conception, which makes us see all relief as a transitory stage in an evolution more or less completed, and transforms into living realities the relief forms that seem to be frozen in appearance. Since the time of Ratzel and P. Vidal de la Blache studies in human geography have manifested a similar tendency, and the data of history are increasingly pressed into service to explain the present state. The historical point of view is less marked in climatology and hydrography; it is already perceptible in botanical and zoological geography. Its application, which is essential in all study of general geography, is perhaps not without danger in studies in regional geography, where description must take first place. The tendency to look at all facts historically is, however that may be, the most striking trait in the evolution of geography in these last years. It is the consequence of an increasingly rigorous application of the principle of causality.

Such are the three essential principles of geographical method, but it is important to understand that the true geographer ought to have all three constantly before his eyes (28). That is something that has not always been sufficiently brought out.

Cartography and geography are not synonymous, and to be a geographer it is not sufficient to depict the extent of any phenomenon whatever. The concern for general laws is a scientific principle; the search for causes is a philosophical preoccupation. But the geographer is the only scholar who forces himself simultaneously to know about the distribution of surface phenomena, physical, biological, or economic, to discern the causes of that distribution, while relating it to the general laws, and to investigate their effects. He is thus led to look at local combinations of influences whose complexity surpasses all that the physicists, botanists, and statisticians imagine. The surface of the earth is his laboratory, a marvelous field of experiences, where are realized an amazing variety of regional types, of which it concerns him to recognize and explain the originality.

What is fruitful and original at the same time in geographical method is that it puts terrestrial realities in view. The genus Quercus is an abstraction; nature shows us forests of oaks, with a whole cortege of associated plants. The extent and the physiognomy of these forests are, in each place, the result of a certain equilibrium among the various and particular influences at that place: climate, soil, relief, exposure, clearings, and crops. Heavy industry is an abstraction; the reality would be the industrial groupings determined by the local combinations of favorable circumstances: presence of coal, transport facilities, and a dense and active population.

In sum, modern geography looks at the distribution on the surface of the earth of physical, biological, and human phenomena, the causes of this distribution, and the local relationships of these phenomena. It has an essential scientific and philosophical character but also a descriptive and realistic character. It is that which makes its originality. Like all the modern sciences, it is under the necessity of calling upon data from various disciplines. If the geologist must be a chemist, physicist, and zoologist, and if the physicist himself cannot do without mathematics, the horizon of the geographer is especially vast and extends at the same time towards the physical, biological, and social sciences.

The development of geography is, therefore, conditioned in a certain measure by that of the auxiliary sciences. A century ago Ritter and Humboldt clearly showed the way; hardly fifty years ago we began to advance it, following Richthofen and Vidal de la Blache. We are still far from knowing all that is necessary for us to understand fully the mechanism of climates, what rules the evolution of relief, and the conditions of the formation of plant associations. The network of meteorological stations is still very loose, and the very laws that govern the movements of the atmosphere are scarcely beginning to be cleared up by the study of the upper layers.   All the ideas on the formation of mountains have recently been controverted by the extension of the theory of drifting.   The very causes that periodically rejuvenate erosion are being discussed.

The necessity of following the changes in point of view of very different sciences is not one of the least difficulties for modern geography.   In specializing, as that will become more and more necessary, the geographer risks losing sight of the very essence of geography, which is to study the interrelations of phenomena.   A look backward will always be the surest remedy.

Few sciences are as old as geography, for it responds to an essential need of the human spirit.   Conceived at first as a catalogue of relative positions on the surface of the earth, it early sought to record the aspects of the earth and of human life.   In proportion to the development of the knowledge of the earth's surface, revealing very different regions, one has seen enriched the detailed descriptions, which, to be understood, demand scientific knowledge.   The necessity of explaining the contrasts and the similarities has led to the idea of laws of general geography. Finally, when the physical and natural sciences have permitted a more complete understanding of local relationships, geography is revealed as a descriptive and explanatory science of a very great complexity and of both philosophical and practical interest.

Such is the route traveled.   It is a matter of continuing in the same direction.   The future of geography could not be more exactly defined.

De Martonne's Bibliography (with emendations)

I.--The Geography of the Ancients

1. Wolfgang Dröber. Kartographie bei den Naturvölkern. (Inaugural dissertation, University of Erlangen, 1903) Erlangen: Buchdruckerei von Junge & Sohn, 1903.

2. Louis Vivien de Saint-Martin. Histoire de la géographie et des découvertes géographiques depuis les temps les plus reculés jusqu'à nos jours. Paris: Hachette, 1873. (Accompanied by atlas published in 1874)

3. Oskar Peschel. Geschichte der Erdkunde bis auf A. v. Humboldt und Carl Ritter. Munich: J. G. Cotta, 1865.

4. Friedrich Ratzel. Die Erde und das Leben. Eine vergleichende Erdkunde. 2 vols. Leipzig and Vienna: Bibliographisches Institut, 1901-1902.

5. Elisée Reclus. L'Homme et la terre. 6 vols. Paris: Librairie Universelle, 1905-1908.

6. Ernst Hugo Berger. Geschichte der wissenschaftlichen Erdkunde der Griechen. 4 vols. Leipzig: Veit & comp., 1887-1893 (2nd ed., 1903).

7. Edward Herbert Bunbury. A History of Ancient Geography, among the Greeks and Romans, from the Earliest Ages till the Fall of the Roman Empire. 2 vols. London: John Murray, 1879 (2nd ed., 1883).

8. François-Amédée Thalamas. La Géographie d'Eratosthène. (Doctoral thesis) Versailles: Imprimerie de C. Barbier, 1921.

9. Joachim Lelewel. Géographie du moyen âge. 4 vols. and atlas. Brussels: Vve et J. Pilliet, 1852-1857.

10. Manoel Francisco de Barros e Jonsa de Mesquita de Macedo Leitão e Carvalhosa, 2nd Vicomte de Santarem (catalogued under Santarem). Essai sur l'histoire de la cosmographie et de la cartographie pendant le moyen âge, et sur les progrès de la géographie après les grandes découvertes du XVe siècle. 3 vols. Paris: Imprimerie de Maulde et Renou, 1848-1852.

11. Konrad Kretschmer. "Die physische Erdkunde im christlichen Mittelalter." Geographische Abhandlungen, vol. 4 (1889).

12. Charles Raymond Beazley. The Dawn of Modern Geography. A History of Exploration and Geographical Science. 3 vols. London: John Murray, 1897-1906.

13. Joachim Lelewel. La Géographie des Arabes. 2 vols. Paris, 1851. [This title does not appear to exist. Perhaps it is part of Lelewel's Géographie du moyen âge (above).]

14. Adolf Erik Nordenskiöld. Periplus, an essay on the early history of charts and sailing-directions. (English translation by Francis A. Bather of Swedish original) Stockholm: P. A. Norstedt & söner, 1897.

15. Paul Vidal de la Blache. Marco Polo, son temps et ses voyages. Paris: Hachette, 1880 (2nd ed., 1891).

16. Sophus Ruge. Geschichte des Zeitalters der Entdeckungen. Berlin: G. Grote, 1881.

17. Adolf Erik Nordenskiöld. Facsimile-Atlas to the early history of cartography with reproductions of the most important maps printed in the XV and XVI centuries. (Translated from Swedish original by Johan Adolf Ekelöf and Clements R. Markham) Stockholm: P. A. Norstedt & soner, 1889.

18. Lucien Gallois. Les Géographes allemands de la Renaissance. Paris: E. Leroux, 1890.

III, IV.—Modern Times

19. Lucien Gallois. "La Géographie générale de Varenius." Journal des savants, n. s. vol. 4 (1906).

20. Christian Sandler. Die Reformation der Kartographie um 1700. Munich: R. Oldenbourg, 1905.

21. Emil Wisotzky. Zeitströmungen in der Geographie. Leipzig: Duncker und Humblot, 1897. (A series of interesting studies for the end of the 18th century and the first half of the 19th.)

V.--The Definition of Geography

22. Ferdinand von Richthofen. Aufgabe und Methoden der heutigen Geographie. Leipzig: Veit & comp., 1883. (One of the first attempts at the definition of modern geography; cf. Triebkräfte und Richtungen der Erdkunde im neunzehten Jahrhundert [Rektoratsrede, Berlin, 1903] [Berlin: Universitäts-buchdruckerei von G. Schade (O. Francke), 1903])

23. Hermann Wagner. "Bericht über die Entwickelung des Studiums und der Methodik der Erdkunde." Geographisches Jahrbuch, vol. 9 (1882), 651-700; vol. 10 (1884), 539-650; vol. 12 (1888), 409-460; vol. 14 (1890/91), 371-462. (Gives an overview of the discussions that have multiplied in Germany on the method and object of geography)

24. Paul Vidal de la Blache. "Le Principe de la géographie générale." Annales de géographie, vol. 5 (1896), 129-142.

25. Alfred Hettner. "Die Entwicklung der Geographie im 19. Jahrhundert." Geographische Zeitschrift, vol. 4 (1898), 305-320. (Cf. "Grundbegriffe und Grundsätze der physischen Geographie," G. Z., vol. 9, no. 1 [1903], 21-40, no. 3, 121-139, and no. 4, 193-213; and "Das Wesen und die Methoden der Geographie," G. Z., vol. 11, no. 10 [1905], 545-564, no. 11, 615-629, and no. 12, 671-686.)

26. Paul Vidal de la Blache. "Leçon d'ouverture du cours de géographie." Annales de géographie, vol. 8, no. 38 (15 March 1899), 97-109.

27. William Morris Davis. "An Inductive Study of the Content of Geography." Bulletin of the American Geographical Society, vol. 38 (1906), 67-84. (Presidential address at the Second Annual Meeting of the Association of American Geographers, New York City, December 1905) Reprinted in William Morris Davis, Geographical Essays ed. by Douglas W. Johnson (Boston: Ginn and Company, 1909), 3-22.

28. Emmanuel de Martonne. "Tendances et avenir de la géographie moderne." Revue de l'Université de Bruxelles, vol. 19, no. 6 (March 1914), 453-479.

THE LESSONS TO DRAW FROM THE HISTORY OF GEOGRAPHY

Paul Claval

(This is a translation of Chapter 1, "Les Enseignements à tirer de l'histoire de la géographie," of Pour le cinquantenaire de la mort de Paul Vidal de la Blache by Paul Claval and Jean-Pierre Nardy [Cahiers de géographie de Besançon, 16] [Paris: Les Belles Lettres, 1968], pp. 11-19.)

Works relating to the history of geography know an unequal favor according to era. In the second half of the 19th century, they were numerous. In France, to be sure, the majority of geographers came from the historical sciences. But the same interest is seen elsewhere, in the Anglo-Saxon countries and in Germany, which indicates that there is something more in the general curiosity shown in problems of evolution than just the simple influence of a neighboring discipline.

The publications of that period accorded a very large place to the narrative of the discovery of the earth and the history of its cartographic representation.[1] The parts relating to geographical thought are less numerous and much less rich. In France, for example, the most successful studies have sometimes dealt with travels of discovery--Vidal de la Blache thus wrote a book on Marco Polo[2]--or with the great cartographers who had made such great progress in the scientific knowledge of the earth since the 14th century: Lucien Gallois devoted his doctoral dissertation to the German cartographers of the Renaissance.[3]

From the beginning of the 20th century, geographers became field men; armchair scholars are increasingly rare. Studies in the history of geography became less numerous, outside of certain specialized areas. They ceased to interest the big names of the time, who were all occupied with multiplying regional studies or systematic syntheses. These studies were often conducted by historians, so that ancient geography is the subject that continues to maintain the most continuous current of thought. Works relating to the discovery of the world are going out of style; they are taken up by popularizers, reaching a large public, but are no longer as highly valued by men of science. The analysis of cartographic productions is the work of erudites, librarians, and museum curators. Hardly anywhere outside Italy can one still find first-rate geographers who devote part of their researches to the analysis of manuscript sources and old maps.[4]

Scientific production relating to the history of geography is less abundant, being more apt to be the work of erudites, but it would be wrong to deduce too quickly that the field is not being changed or enriched and that it continues unceasingly to employ the same procedures and to be interested in the same things. Several works show the progressive enlargement of curiosity. More and more one sees in the history of geography something other than a narrative of the discovery of the world and of the invention of methods of representation that permit the organization of results. In Germany and the United States writers turned resolutely to problems in the conception of geography; Hettner[5] on the one hand and Hartshorne[6] on the other saw in the study of the evolution of geography a means of advancing epistemological reflection on the nature and objectives of our discipline. John Kirtland Wright[7] opened other perspectives: beyond the geographical knowledge of particular phenomena that is done in laboratories, in the offices of men of science, or, formerly, in the great adventures of exploration, there is a general geographical knowledge that characterizes a period, explains it in part, but is explained by it. Wright shows us how the world views were organized in the time of the Crusades and makes us understand that it concerns a world system that demands attention. Its logical structures are organized according to the general philosophy of the time. What does it matter if the details are wrong, or if the ideas are more in the realm of myth than in that of rational study? An error can have as much effect on the plan of action as a truth. It informs quite as well, if not better, on the general context of knowledge of a period. The history of geography thus assumes a new dimension; it becomes cultural or social, and shows us the relativity of the great systems that are imposed in this or that period. Father de Dainville provides us an analogous lesson on another plane. His Géographie des humanistes[8] has an ambition apparently less large than the study that Wright had dedicated to the geographical knowledge of the time of the Crusades, in the sense that it treats a narrower theme. How, in the ensemble of disciplines taught by the Jesuits, was a place made for geography? What are its relations with the ensemble of philosophical positions that define their attitude? Those are the questions that Father de Dainville poses. History thus reconstructed is fascinating: one follows in it the enrichment of a science born of scholasticism, but which little by little integrates what the published accounts of the mission fathers have done to illumine the world. A perpetual dialogue is established among the particular sciences that permits the shaping of changing intellectual atmospheres and brings about or necessitates a progressive transformation of the initial premises of each discipline. The history of a science thus opens on that of all knowledge, on the whole vision of the world at one period.

For twenty years, works relating to the evolution of geography have multiplied anew. In Germany certain authors have tried to define its place in the totality of research activities.[9] Elsewhere, in England, the United States, and more recently in France, the intellectual trend is parallel, although the research goals are less systematically defined and the works are done in a more diffuse and less systematic manner. But the studies present many analogies from one country to another. They bear on the history of geographical thought and often are registered in the same line as those of Hartshorne. Historical method makes constant progress, which is shown for example in the publication of ancient texts and in the systematic reprinting of old atlases and travel journals.[10] Biographies bring back to life personalities who were important but who were only imperfectly known. In Germany, Alexander von Humboldt and Carl Ritter have attracted numerous researchers, so that there are few aspects of their work that are today not well known.[11] In America, George Perkins Marsh[12] has been rediscovered and, through him, the origins of all the conservation philosophies. Geographical journals give a large place to reviews and notes, which cause refinements on particular aspects of geographical thought to be multiplied.

The evolution of the totality of studies in the history of geography, their favor more or less great according to the times, and their changing nature are explained rather easily. When geography finished getting established as a scientific discipline at the end of the last century, the need to draw up an inventory of what had already been written was pressing. What geographers had succeeded in doing, in the course of the preceding periods, was to erase the majority of the blank spots on the map, the zones of terra incognita that had long monopolized attention. Bourguignon d'Anville, in the 18th century, in allowing lacunae to remain on the maps where data were lacking to him, had clearly defined what seemed essential in order to be able to perfect the knowledge of the configuration of the lands and seas. At the end of the 19th century the movement of exploration reached its end; there remained to reconnoitre only part of the polar lands, certain high parts of the great mountain masses, and the essential elements of the topography of the ocean basins. The interest of certain of the great geographical societies in the problems of exploration remains active to the present day. That is the case, for example, with the Royal Geographical Society. But the problem is no longer at the heart of scientific preoccupations. The progress of the whole of knowledge no longer depends on that extension of the limits of the known world that had long obsessed the previous generations. The history of the discovery of the world continues to fascinate a large number of people, but it is found on the margins of the great movements of geographical thought, which explains how it has become, for the most part, a domain given over to erudites, who are often indifferent to the general growth of the body of knowledge of the discipline.

One also realizes that the history of discovery and of the representation of the earth, which interested geographers almost exclusively a short time ago, is marked by a certain ethnocentrism; if one knows how the civilizations that have formed western culture have progressively strengthened their knowledge of the earth, one often ignores what the other peoples have learned. This is an area for some fundamental spade-work; progress in it has been rapid for a generation. We are starting to get a precise idea of the geography of the Arabs.[13] The peoples of the Far East have undoubtedly equalled them, and even if they had not had such a great curiosity for foreign lands, one cannot ignore the progress of geographical exploration accomplished by the Chinese[14] in the course of what corresponds for us to the Middle Ages and to the beginning of the modern period. One begins to see more clearly today that the data of the knowledge of the earth are not absolute, that there have been as many explorations and curiosities as there are great cultural domains or great civilizations. That is a lesson that is not a bad one to give to the students we are preparing; we are too willingly inclined to ethnocentrism in scientific matters.

At the end of the last century geography was constituted to go beyond simple topographic description and the simple recitation of exploration. It wanted to be explanatory, and it took more and more consciousness of the diversity of the characteristics of the earth's surface. It had to devise new methods and to address problems that the researchers of the previous generations had often neglected. Close study would have shown that the effort of explanation and systematization was already considerable with many of the writers of the 17th or 18th centuries. Nevertheless, one can well understand the myopia of the majority of geographers at the end of the last century; they had the feeling that they were founding a new discipline. They found few guides to aid them in solving their problems. Outside of Ritter and Humboldt, they recognized hardly any predecessors.

With the passage of time, one sees that their effort is less remarkable than they thought; their construction had not been made starting from nothing. They freely used the work of travelers, explorers, and compilers that had suggested certain of their approaches to them. They had been steeped in an intellectual atmosphere without which one could not understand certain of their prejudices or certain of their more fruitful intuitions. One becomes more aware of the diversity of attainments and methods of the discipline that was formed in the course of the last decades of the past century. The study of geographical thought thus takes on interest. One sees how it is tied to the whole evolution of scientific thought at the time. One can also see that the scientific methods and theoretical constructions cannot always be understood without reference to the historical framework in which they were thought up. As time passes and researchers are multiplied, one perceives more clearly the scope and significance of the development of geography, as it is now analyzed.

History has been taught for a long time because it offers useful lessons to shape youthful minds or fashion their characters. One makes fun of narratives in the style of Plutarch, but it is certain that one is not entirely right in doing so, for the pedagogical virtue of reconstructions of the past is rich and very diverse.

First of all, the history of geography provides us with lessons of relativity and scientific modesty; none should ignore them. The scholar is never an isolate; he is profoundly plunged into a milieu and into an era. He sometimes goes ahead of it and anticipates what tomorrow's scholarship will be like. But his experience is never registered completely marginal to the general social context. On the whole, scientific advances are made more as a succession of little results than of sharp and triumphant changes.[15] Specialties often develop along parallel courses in ignorance of each other. In the domain of the human sciences, in particular, analogies are striking, and one cannot comprehend why so many researchers refuse to recognize the common points, the similarities of method, and even sometimes the identity of the goals of research.

For the geographer, the first lesson is of modesty. It is also the first lesson for the discipline as a whole, because it cannot make a stand against others like a great solitary figure. Geography finds herself tied by all her traits to that which is done around her. But she profoundly marks the milieu that receives her. That is something that often escapes those whose academic formation has removed them from the action. They have the impression of living outside of time, of being without a hold on their epoch. They often express a certain bitterness for it and secretly envy those whose art more easily or directly finds a buyer. Many analyses show the inexactness of these impressions. The ways in which scientific geography exercises an influence on its epoch are often indirect. It is by the press, publications, and teaching that certain facts find an echo in a large milieu and contribute to change the world. The demonstration of it has been made for a long time for certain moments in history. Columbus's venture is inseparable from all the history of geographical thought in the 15th century; he sailed because he believed in the sphericity of the earth. He embarked lightheartedly, for the calculations of the scholars of the time were wrong and Asia seemed nearer by the western route, which proves that error can be as effective as truth in the domain of the influence of a science on the course of history! Recent studies place in evidence analogous cases, and the demonstration is easier to make of the influence of an idea or a theory when its scientific basis is more fragile. Henry Nash Smith[16] has written an essay on the role of the West in American life. It is not the work of a professional geographer; the author is a specialist in the history of American literature, but the theme of his quest is geographical in so many of its traits that it is difficult not to see in the essay one of the most original contributions to the history of the relations of academic geography, popular knowledge, and the great political decisions. The West has been one of the most powerful myths in American history. It has acted much more by the manner in which it was conceived and constructed than by any inherent attraction. The myth is nourished by historical reminiscences and by old dreams that extended those of Columbus. The western quest had its origin as a passage toward the East, of a new way permitting the rediscovery, by shorter routes, of the great traditional centers of world civilization. Curiously, this quest had drawn certain of its justifications from the climatological works of Humboldt. The theory of the isothermal belt, the zone that by the qualities of its climate offers the best chances for the flowering of humanity, has reinforced the faith that many place on the value of the West for the American homeland. It is because one believed in the general migration of centers of civilization along that isothermal belt, from East to West, according to a kind of general law of the historical development of humanity that the cult of the virgin wildernesses that extend beyond the frontier often has messianic aspects. The strength of geographical myths appeared such that John Wesley Powell[17] had difficulty destroying them. Many prefer the dream that clings to the old conceptions to the prosaic reality of these hard and mostly poor lands. The experience is full of significance for the history of geography, for it was in part to destroy the illusions nourished by a badly understood geography that Powell created modern geography, which has caused the scientific method to triumph in the United States.

John Kirtland Wright[18] gives us other examples of that power of geographical thought in the determination of the currents of history. The false notions fascinated him particularly, for they allowed him to show the surprises of history and the often unsuspected richness of that which is controverted by the facts. The dream of a navigation route by the Northwest, between Europe and the Far East, progressively vanished since the 17th century. All the voyages undertaken showed the presence of barriers of ice-floes, forbidding navigation beyond Hudson Bay and Baffin Bay. Then some scholars put forth the hypothesis of a polar sea free of ice[19]; the pack-ice did not extend up to the pole, and if ever one happened to find a break in its front, one could have the benefit of an open sea, which would assure the link by the shortest route among the lands of the Arctic basin. Suddenly expeditions in the direction of the northern lands multiplied anew. They brought about the destruction of the myth of the open sea, but they opened the way to the systematic exploration of the polar lands and brought about considerable progress in the oceanography of those regions.

It might be feared that habitual reading of studies in the history of geography gives an inexact view of the respective role of precise studies and the fascinating but sometimes vain ideas. Is it good to show the immorality of scientific evolution and to see error better compensated than truth? The danger is very slight, and the well done works show the sometimes modest origin of essential progress. The history of geography is not simply that of doctrines, theories, or hypotheses. It is also that of techniques; their role in the progress of thought is such that one sees that it is impossible to dissociate the analyses of epistemological systems from those of the ways of the time in which they were developed. Father de Dainville announced, in his Géographie des humanistes, that he would devote a later study to the scientific vocabulary of the classical age. He waited more than twenty years to put the finishing touches on this Langage des géographes,[20] which is without doubt the most interesting and the freshest work that has been published in France since the war in the domain of the study of the history of our discipline. One sees in it the patient work of that unjustly forgotten world of the engravers and cartographers; unceasingly, they stated precisely what they were portraying and shaped the vocabulary of geographers while depicting the various landscapes. What seems to us today so simple and natural, the vocabulary of topographic description, for example, could take shape only because of that daily confrontation with the physical reality that one seeks to translate with the greatest possible fidelity. One rediscovers, on reading their work, a whole world that is today gone, that of multiple administrative districts, ecclesiastical seats, and juridical jurisdictions, and one finds out about the evolution of many of the essential concepts, that of climate, for example, which the topographers did not define, but of which they were quick to use the new meaning that the medical doctors gave it and that we know exclusively. Beyond graphic problems, of that of outline and clarity, two centuries of geographical work appear. The artists were not responsible for everything. They must not cause us to forget the parallel researches of those humanists whose history has been written elsewhere, but the general progress of the science of geographers is as much the work of cartographers, of those who struggle with the technical problems, as it is that of men of pure thinking.

Under one aspect or another, it is always a lesson of relativity that we get from the study of the development of geography, of its influence, or of its place among the sciences. That goes much farther than one generally thinks. When one teaches a discipline without placing it in a historical perspective, one has a tendency to give it a coherent, clear, and logical image. The tableau lacks depth and shading. One has the impression that all parts are equally solid and equally necessary to the harmonious equilibrium of the whole. It naturally happens that one gets a feeling of annoyance and a certain malaise in the presence of statements that seem less narrow and less evident than others. When one has a certain habit of critical work, one concentrates quickly on these points and explores them with a very particular attention, for one understands that it is there that the edifice is less solid and will collapse if it is not rebuilt in time. But many forces hold back when one finds out the imperfections of the general construction. Should one rely on that somewhat uneasy feeling that one feels in his conscience? Is it the science that is defective, badly done, and badly structured? Has one well thought out all the elements of the problem? Is the logical coherence profound or does it come solely from an imperfect account of the scientific theory at the time? Isn't it simply because certain stages of reasoning are missing that one thus remains restless and unsatisfied? These are the questions that one quite naturally asks himself and that discourage a criticism in depth. Moreover, very often, the habit of seeing things set forth in a certain order is so old and so deeply anchored in the reflexes of the geographer that he is accustomed to it and no longer perceives so clearly the logical flaws and the incoherences of the statement. In the majority of studies in human geography, what they try to analyze are complex economic functions and activities in which the division of labor implies the specialization of each in different branches. The manuals that ought to prepare, on a general plan, an analysis of the facts of human geography, usually give only a modest place to these forms of organization. On the contrary, they dwell at length on all that concerns the analysis of genres de vie, which is no longer practiced. Why? Is the tableau of human geography then incoherent? At first we think so, but then we have second thoughts; isn't there some reason of high order that one momentarily overlooks to account for the striking hiatus between the science that is practiced and that which one depicts?

When one puts problems back into their historical development, the judgments become easier. Progress is often made by circuitous routes, so that the science that is formed can never present itself under the form of a perfectly unified and coherent construction. Certain interpretations lag behind others, and certain parts are transformed, while others remain faithful to the most ancient recipes. The effort that one constantly goes through to put the diverse elements in their proper relations is necessary. It is fruitful, for without this need of coherence and the profound appeal of logic, the front line of progress would be still more irregular and the divergences more marked from one field to another. But in the moments of trouble and doubt, historical analysis becomes one of the fundamental tools of research. It shows in what general context such and such a theory is elaborated. It permits one to understand their deep significance and to find out the higher reasons that analysis of the present cannot yield. It shows also that the elements are relative and that they ought to be constantly reinterpreted, for the scientific language of a period becomes obsolete. Should one neglect the study of genres de vie? Yes, apparently, if one limits himself to studying the geography such as it is lived today. Yes, also, if one studies the way in which they have appeared one moment as one of the essential pieces of the geographical edifice, of which they have summed up, for a generation, a certain harmonious manner of forming the place of human societies in the world. But at the same time one understands that they play a double role in the general structure of science. They describe certain ways that the societies have of taking advantage of the environment in which they are found, but permitting also to show the social nature of human geography and to clear up certain fundamental aspects of the structure of the science. Nothing justifies the exclusive attention that had long been accorded to them by the theoreticians, if they are really insufficient to describe modern societies and the manner in which they are tied to the environment that makes them live. But the historical perspective shows that in touching one stone, the whole construction is implicated, and that it is necessary, if one tries to rethink that aspect of the classical doctrine, to take up again also the whole edifice that rests on it.

The historical method is thus one of the favored instruments of all thought on the development of science, of all effort of logical construction, and of axiomatic analysis, one could almost say. It facilitates criticism and often justifies the intuitions that one has had in contemplating the present equilibrium of the general theoretical construction. Each time that one tries to touch one of its elements, it shows what its role is in the living structure of the whole. In that sense, historical analysis guides fundamental research, since it makes it discover the links that simple superficial examination cannot reveal, for they are illuminated only by a now forgotten context.

The study of the history of geography is sometimes still considered as a pastime of the erudite, as a way that is opened to professors near their retirement to retrace their past activity and transmit to the succeeding generations the memory of what they have experienced. That is to see only a small part of its interest. On the pedagogical level, it offers an occasion, difficult to find elsewhere, to be initiated into textual criticism, one of the fundamental steps in all scientific activity but one which very few practice in our discipline. On a more general level, it seems to us fundamental, for it alone permits the constant reevaluation of the objectives and methods of the scientific quest, without breaking with tradition and without losing the lessons of the preceding generations. It permits the making of a geography that is always new and always coherent but that doesn't deny its origins and knows how to derive profit from the thoughts of past generations.

Claval's Footnotes (with emendations)

1. For example, one can refer to Louis Vivien de Saint-Martin, Histoire de la géographie et des découvertes géographiques depuis les temps les plus reculés jusqu'à nos jours, Paris: Hachette, 1873 (accompanied by atlas published in 1874).

2. Paul Vidal de la Blache, Marco Polo. Son temps et ses voyages (Paris: Hachette, 1880).

3. Lucien Gallois, Les Géographes allemands de la Renaissance (Paris: E. Leroux, 1890) (reprinted Amsterdam: Meridian Publishing Company, 1963).

4. As witness the work of Roberto Almagià. [Almagià (1884-1962) was Professor of Geography in the University of Rome, 1915-1958]

5. Alfred Hettner, Die Geographie, Ihre Geschichte, Ihr Wesen, und Ihre Methoden (Breslau: Ferdinand Hirt, 1927).

6. Richard Hartshorne, "The Nature of Geography. A Critical Survey of Current Thought in the Light of the Past," Annals of the Association of American Geographers, vol. 29 (1939), 171-658 [Also issued in book form by the Association and kept constantly in print]

7. John Kirtland Wright, The Geographical Lore of the Time of the Crusades. A Study in the History of Medieval Science and Tradition in Western Europe (American Geographical Society, Research Series, No. 15) (New York: American Geographical Society, 1925) (Reprinted with a new introduction by Clarence J. Glacken, New York: Dover Publications, Inc., 1965).

8. François de Dainville, La Géographie des humanistes (Paris: Beauchesne, 1940).

9. For example, one can consult on this subject the following papers by Hanno Beck: "Methoden und Aufgaben des Geschichte der Geographie," Erdkunde, vol. 8 (1954), 51-57; "Entdeckungsgeschichte und geographische Disziplinhistorie," Ibid., vol. 9 (1955), 197-204; and "Geographiegeschichtliche Ansichten," Geographische Zeitschrift, vol. 55 (1957), 81-90.

10. In England the publication of travel narratives has been systematically pursued since 1846 by the Hakluyt Society. The work has been carried on there with a continuity much greater than everywhere else, but there has not been an appreciable acceleration of researches for a decade. Among the latest [i.e., by 1968] texts published, let us cite The Principall Navigations, Voiages and

Discoveries of the English Nation by Richard Hakluyt [2-volume reprint of 1589 edition in 1965] and the travels of Ibn Battuta [3 vols. edited by H. A. R. Gibb, 1958-1971; 4th volume (edited by C. F. Beckingham) not yet published].

In Germany, a considerable effort is made to assure the publication of travel narratives, thus to better understand the biography of great geographers.

In the Low Countries, numerous old texts have recently been republished, as well as a very rich collection of maps and atlases of the 16th, 17th, and 18th centuries (in the collection of Theatrum Orbis Terrarum [a publishing house in Amsterdam]).

11. One can find information about studies relating to these two authors in our Essai sur l'évolution de la géographie humaine (Paris: Les Belles Lettres, 1964), pp. 27-29.

12. It seems that it was Lewis Mumford who first recalled the significance of the work of George Perkins Marsh. The most complete study on this author is David Lowenthal, George Perkins Marsh: Versatile Vermonter (New York: Columbia University Press, 1958).

[Some explanation is required for the first sentence in this note. Although he had never been completely forgotten, the modern popularity of Marsh's work among American geographers seems to derive from the interest of Carl Sauer (1889-1975) and his students in the University of California, Berkeley. Sauer was introduced to Marsh's book by a Dutch exchange student, Jan O. M. Broek, in the summer of 1931. Broek had bought a copy of the second edition on 11 July 1931. In November 1931, on his return to Holland, Broek purchased a copy of Lewis Mumford's The Brown Decades in New York and noticed a reference to Marsh. (Letter from J. O. M. Broek to G. S. Dunbar, 31 July 1966.) Mumford and Sauer teamed up in 1955 to lead (with Marston Bates) the Wenner-Gren Symposium, "Man's Role in Changing the Face of the Earth," which was referred to as a "Marshfest" because G. P. M. was the acknowledged guiding spirit of the conference. The resulting book, Man's Role in Changing the Face of the Earth (Chicago: University of Chicago Press, 1956), was dedicated to Marsh. It was Carl Sauer who first introduced David Lowenthal to Marsh's work.]

13. One can find a short account of the knowledge in this area in A. Torayah Sharaf, A Short History of Geographical Discovery (Alexandria, Egypt: M. Zaki el Mahdy, 1963) (Reprinted London: Harrap, 1967).

14. René Grousset (1885-1952) has shown the French public certain aspects of this discovery of the world by the Chinese in retracing the adventures of Buddhist pilgrims who reached India in the 6th and 7th centuries. [Sur les traces du Bouddha (Paris: Plon, 1929), translated into English by Mariette Leon as In the Footsteps of the Buddha (London: G. Routledge & Sons, 1932, reissued 1971)]

32

15. On this point, one can consult the little book of Kourganoff; it does a good job of dissipating the romanticism that generally surrounds all evocations of scientific research. Vladimir Kourganoff, _La Recherche scientifique_ (Collection "Que sais-je?," No. 781) (Paris: Presses Universitaires de France, 1958, and subsequent editions).

16. Henry Nash Smith, _Terres vierges_ (Paris: Seghers, 1967), French translation of _Virgin Land: The American West as Symbol and Myth_ (Cambridge, Mass.: Harvard University Press, 1950).

17. On this point one can consult R. J. Chorley, A. J. Dunn, and R. P. Beckinsale, _The History of the Study of Landforms_, Vol. 1, _Geomorphology before Davis_ (London: Methuen, 1964), pp. 469-602.

18. John Kirtland Wright, _Human Nature in Geography_ (Cambridge: Harvard University Press, 1966).

19. _Ibid_., Chapter 6, "The Open Polar Sea," 89-118. Among the defenders of the idea of the open polar sea, Matthew Fontaine Maury (1806-1873) was undoubtedly the one whose authority was greatest in the United States.

20. François de Dainville, _Le Langage des géographes_ (Paris: Editions A. et J. Picard et Cie, 1964).

GEOGRAPHY: THE HISTORY OF GEOGRAPHY, CHRONOLOGICAL EVOLUTION, AND TRENDS IN GEOGRAPHICAL THOUGHT

Philippe Pinchemel

(This is a translation of Pinchemel's article, "Géographie: l'histoire de la géographie, évolution chronologique, les tendances de la pensée géographique," in Encyclopedia Universalis, vol. 7 [Paris, 1970],pp. 621-625.)

A. The History of Geography

The history of geography is not the history of a science seen through its developments, its "great men," and its methods, but a world of questions to pose and choices to make.

For centuries, some men have been considered or have considered themselves geographers, whose professions and works are so different that the historian is right in asking himself about their belonging to the same field: Herodotus and Montesquieu, Eratosthenes and Kant, as well as the pedagogues, the "geographers to the king" who were charged with describing battlefields and the sieges of cities, and the geographer-engineers who made the maps.

The history of geography is easily assimilated to the history of discovery and the history of the knowledge of the earth; was not the geographer and is he not, then, the explorer, the discoverer of new worlds, the lecturer returned from distant lands, or the one who, often without traveling, compiles valuable geographical dictionaries or edits useful manuals? It is really the history of a science as old as the earth, responding to one of the first questions of Man: Where am I? The history of geography also retraces the evolution of one of the most fascinating and complete disciplines that there is, which is constantly renewed, always questioning, and thus always young.

This uncertainty about its nature and its object doubtless explains the extreme discretion of geography about its history, but the tendency seems to be changing with the creation of a Commission on the History of

Geographical Thought in the International Geographical Union (1968) and the existence, in several countries, of specialized research centers.

An Evolution Tied to the Advances of the Sciences

The history of geography is the history of Man's perception of his environment: it shows exactly how, over the ages, men have become aware of the diversity of the terrestrial globe, the ideas and interpretations drawn from that awareness, and the successive theories that came from them; it studies the ways in which Man, embodied in societies that are themselves diverse, situates himself with respect to his habitat; it recognizes in what ways Man has given significance to the coherences or incoherences of the natural world. The perception, for example, of the fact of "valley," the organization of two slopes on both sides of a line of low points that follows a watercourse, was not made at the dawn of humanity, still less that of the causes of the existence of valleys.[1] It is tied to the evolution of "the language of geographers,"[2] to the birth of new words, an expression of the growing awareness of new concepts and new geographical entities.

In fact, the history of geography is inseparable from all history. Even more than is the case with other sciences, geographers cannot be "extracted" from the environment and epoch in which they have lived. From one century to another not only has the raw material depending on geographical observation changed, but the possibilities of approaching that geographical reality and of understanding it in its various meanings have greatly changed; thus there is little resemblance between geographies without maps and those that make use of them; the history of cartography is not distinct from the history of geography. Before geodetic triangulation and precise leveling, all spatial properties tied to localization, distances, and relief were perceived relatively. The inequality of the state of advancement of cartographic works in the world has inhibited and still inhibits comparative studies. It explains why the geographers of different countries have not worked at the same scales and why the knowledge of parts of the globe is not identical, for the diversity of cartographic documents directs the work of the geographer. It is the same for statistical material; population censuses have played a decisive role in the development of population geography and, more generally, in the orientation and research themes of all human geography.

The Diversification of Tendencies According to Country

The history of geography cannot be dissociated from political, economic, social, and intellectual history and from the evolution of the other natural, economic, and social sciences. French geography, since the 1870s, has been closely tied to colonial history, to primary and secondary teaching, and to Darwinian thought. Between the two World Wars its orientation was influenced by the philosophy of Henri Bergson.[3]

In Eastern Europe and the Soviet Union geography is strongly marked in its research directions and interpretations by Marxism and by Communist ways of thinking. British geography is inseparable from the industrial and urban revolution, from maritime, colonial, and capitalist expansion, and from the preeminence of economics. The geography of the United States, appearing tardily on the scene, is indissociable from the westward movement of settlement as shown by the expression of a frontier of a very great fecundity.

In retracing the history of geography, two series of factors appear to be major:

1. The important role of the environment and mode of life in the shaping of thought and in the elaboration of concepts and judgments. Who, in effect, more than a geographer can be influenced or "determined" by his milieu and by the comparison of successive milieux? The "images of the geographer" and the perception of relative contrasts, distances, and scales are not the same for Brazilian and Italian geographers, or for Canadian and Swiss; likewise, topographic and climatic contrasts and the notions of resource potentialities are not perceived similarly.

2. The no less important role of temperament. Every geographer expresses his personality in the research themes he chooses, in the way he approaches them, and in the interpretations he draws from them. Because it is at the same time a natural science and a human science, geography is more subjective, more diverse, and undoubtedly richer in the variety of temperaments of those who practice it. Geography thus presents multiple aspects: materialistic or spiritual, inductive or deductive, more naturalistic or more human, more analytical or more synthetic, more empirical or more theoretical.

B. Chronological Evolution

To sketch a tableau of the history of geography supposes on the part of the author a personal engagement over what is for him the geography of which he is going to trace the conception--in various forms--from one century to another. A frequent attitude and one that greatly reduces difficulties consists in having geography born only in the 19th century. It seems, however, difficult to forget the innumerable works that by their titles are referred to geography, and to neglect the lives of scholars who called themselves geographers, even if they do not correspond to the subjective image that each geographer has of his science. Above all, one can state that in all eras, men held in their hands the strands that led to a modern conception of geography and perceived what was a true geographical science.

From the Theories of Antiquity to the Arab Discoveries

The balance sheet of ancient geography is not negligible, and its influence extended up to the 16th century. Three currents were formed in the classical period. The first, represented by Eratosthenes, is a mathematical and astronomical geography. Eratosthenes (c. 284-192 B.C.) calculated the circumference of the earth, making conspicuous an order of geometric representation. He divided the earth into "climates" defined by the length of the longest day of the year. His work was followed up by Claudius Ptolemaeus (c. A.D. 90-168), whose cartographic undertaking, Imago Mundi, was the basis of all representation of the earth for centuries; an index gave the coordinates of all known places. From that time on, the idea of climatic (thermal) zones advanced.

The second trend, more historical and descriptive, had for its "leading men" Herodotus (c. 484-425 B.C.), whose travels abounded in pertinent descriptions and explanations, and Strabo (64 B.C.-A.D. 36), whose Geographia in 17 volumes is the first of the geographical dictionaries and encyclopedias.

The third current is embodied by Aristotle and the Stoics, inheritors of Ionian philosophy; phenomena were studied by them according to the four elements of matter. The treatment of Meteors by Aristotle is distinguished by the search for interpretations.

But apart from this last current, ancient geography was made up essentially of measurements, itineraries, and catalogues. In general, it was dispersed among works that were not called geography. Geography was both cosmography and chorography.

During the Middle Ages, the torch of geography left Christian Europe to be taken up again by an extraordinary flowering of Arab geographers, among whom one can cite Obeid Allah Yakut (574-626), the traveler Ibn Battuta (1304-1368), and the geographer Ibn Khaldun (1332-1406). The works of the last-named constitute an intelligent description of the zonal geography of the Northern Hemisphere. But their works had a weak influence on European thought, which continued to live largely on its inheritance from Antiquity. (The works of Ptolemy were translated into Latin in 1416 and printed in 1475; those of Strabo were published in 1516.)

Medieval geography is identified still with cosmology or cosmography. It exists in the form of Imago Mundi, mirrors of nature, and encyclopedic descriptions. Emerging from that period were only rare works, such as the cosmography of an anonymous geographer of Ravenna (650) and the confrontations between neptunists and plutonists.

The Renaissance: 16th-18th Centuries

European geography found the elements of its rebirth in the period of the great discoveries, followed by colonial, commercial, and missionary expansion.

The progress of cartography (the portolans appeared in the 13th century) with mappaemundi, planispheres (maps of Gerard Mercator, 1512-1594), atlases (Theatrum Orbis Terrarum of Abraham Ortelius, 1570, Atlas of Mercator, 1595), travel accounts, and letters and reports from colonial authorities and missionaries produced a realistic vision of the world and brought a new organization of knowledge.

It was the occasion for discussion between the Ancients and the Moderns, for the old conceptions collapsed before the multiplicity of facts. The Cosmographia Universalis (1544) of Sebastian Münster (1488-1542) attempted a systematic description of the world. In La République, Jean Bodin ·(1530-1596) determined the great geographical zones of the globe. Richard Hakluyt (1552-1616) offered at Oxford, in 1574, geographical instruction of an especially cartographic nature and revealed himself as the first great Elizabethan geographer.

Philipp Cluver (1580-1622), of Danzig, wrote an Introductio in universam geographam, whose success was considerable (1624, 1652).

The Dutchman Bernhardus Varenius (1622-1650) gave to that geography its logical framework. His Geographia Generalis (1650) distinguishes between general geography, on the one hand, physical geography, susceptible to generalizations and to scientific observations leading to laws, and, on the other hand, special geography, which, considering societies and their enterprises, is less amenable to scientific method. It is a scholarly work, which makes much more use of theory than of observation, but which traces the limits of a general geography, especially physical.

Aristotelian geography evidently resisted evolution longer, up to the treatment of Meteors, then to Principia philosophiae (1644) of Descartes, a mixture of still ancient theories and of the ingenious intuitions of modern theories.

In the 16th and 17th centuries, modern geography, still very much impregnated with humanism, took form also in teaching, particularly in the schools of the Jesuit Fathers.

The second half of the 17th century and the 18th century represent a paradoxical stage in the evolution of geography. That period of apparent "dilution" of this discipline, without great names and without major publications, is at the same time a period of extraordinary preparation for modern geography as a result of scientific and philosophical development. Research travels with multidisciplinary scientific equipment replace the voyages of pure exploration; the natural sciences are individualized: geology, botany, and agronomy. Considerable cartographic material is produced thanks to progress in cartographic techniques; contour lines appear (1728) subsequent to representations by hachures (1676); the map of France called "Cassini" begins to be published in 1769. Philippe Buache (1700-1773), the "inventor" of drainage basins, published an Atlas physique in 1756; in England the Ordnance Survey was created in 1791.

Above all, the interest in nature and in the relations between nature and society introduced into literature and philosophy what will much later be called a geographical spirit. In various forms the contributions of Buffon, Diderot, Montesquieu, and Rousseau are discernible.

In the framework of that pre-modern geography there developed a natural or physical geography and a human geography preoccupied with the relations between man and nature and with the differentiation of the earth into regions. But this science was still strongly impregnated with the past; at least that is the impression that emerges from the very traditional tenor of the article "Géographie" of the Encyclopédie (1757) signed by Robert de Vaugondi, "Geographer to the King."

The degree of maturity that it reached varied from one European country to another as witness the remarkable works of the German geographers of the 18th century (J. R. Forster [1729-1798] and J. C. Gatterer [1727-1799]). In the Russian Empire, the Moscow Academy of Sciences in 1734 created a section of geography that was directed by M. V. Lomonosov (1711-1765) after 1758.

Kant (1724-1804) participated directly in the evolution of geography. Professor of physical geography at Königsberg, he proposed a definition of geography or, more exactly, of the geographical sciences; compared with the systematic and historical sciences, they studied facts in their spatial relations, their localizations, and their extensions. For Kant, physical geography is the basis of all geography.

The end of the 18th century saw the future of the science more clearly delineated; its progress is especially tied to that of geology. The Britishers James Hutton (1726-1797) and John Playfair (1748-1819) were the authors of theories of the Earth and of the folding of mountains. In the same era, the Russian writer A. N. Radishchev (1749-1802) published his Journey from St. Petersburg to Moscow (1790).

Advances in the 19th Century

1800-1860: Humboldt and Ritter

The first half of the 19th century is a period in which strong personalities appeared: two of them are often held as having played a major role in the evolution of geography, both of them Germans, Humboldt and Ritter.

Alexander von Humboldt (1769-1859) was a geologist, naturalist, great traveler, and prolific author; his work, characterized by its originality and its power of synthesis, had scarcely any influence at the time; later it was considered as eminently "geographical" by its concern for relationships, comparisons, and causes.

Carl Ritter (1779-1859), journalist, armchair geographer, philosopher, spiritualist, and professor, gave his numerous works a human orientation and his scientific spirit led him to important attempts at classification. With Ritter, geography became especially comparative, correlative, and regional.

But, beside them, it would be unjust not to mention other names. Conrad Malte-Brun (1775-1826), Danish by birth but French by adoption, founded the Paris Geographical Society; from 1810 to 1829 he published a Précis de géographie universelle in 8 volumes. Francis Galton (1822-1911), "geographer of the Victorian era," was a specialist in the nascent field of meteorology. The Russian Arsen'ev (1789-1865), a liberal regional geographer, whose Short Universal Geography (1818) went through twenty editions, was the author of essential works such as the explication of the Carte géologique générale de la France (Armand Petit-Dufrénoy and Elie de Beaumont, 1841).

During that same period, the European nations pursued their colonial expansion and equipped themselves with maps, atlases, censuses, and statistical inventories. University teaching of geography made its appearance (Paris, 1809), but the first chairs were still narrowly tied to history and cartography.

The foundation of geographical societies illustrates this rapid institutionalization: Paris (1821), Berlin (1828), London (1830), Russia (1845), and New York (1852).

After 1860:

Ratzel and Vidal de la Blache

Contemporary geography took a more finished form in western Europe in the decades 1860-1900; two men embody this genesis: Friedrich Ratzel and Paul Vidal de la Blache.

The coincidence with an epoch of intense evolution of ideas is striking; in 1859 Darwin published his Origin of Species, underlining the importance of adaptation to the environment. Moreover, the political, economic, and social sciences were being organized; the influence of savants such as Auguste Comte, the Count of Saint-Simon, Frédéric Le Play, Emile Durkheim, and Karl Marx had not been clearly defined but they could not be denied; they brought geography back into balance by developing its social and economic aspects. Furthermore, military and scientific missions and exploration reduced the blank spots in the atlases, leaving to the unknown only the continental "hearts."

Friedrich Ratzel (1844-1904), a naturalist, then journalist and traveler, presented a thesis on Chinese emigration before teaching at Munich (1876) and then at Leipzig (1886). Among numerous publications his essential work was Anthropogeographie (1882-1891). He contributed to "reestablish in geography the human element, whose claims seemed to be forgotten, and to reconstruct the unity of geography on the base of nature and of life" (Vidal de la Blache). For him, knowledge of the Puritan immigrants was more important than that of physical relief in order to "understand" New England. He was a theoretician of space and of place; he would be reproached later for a too-passive image of societies ("People must live on the land that they got by chance, they must die there, and must submit to its laws"); in his writings would be found the embryo of a political geography whose theories were utilized by the supporters of Nazism.

Paul Vidal de la Blache

An historian by training, Vidal de la Blache (1845-1918) "became" a geographer when he took up his first post in higher education at Nancy. His geographical work began in the 1890s.[4] He was truly the founder of the French geographical school, laying the bases of systematic human geography (posthumous edition of Principes de géographie humaine, 1922) and of regional geography (Tableau de la géographie de la France, volume 1 of Lavisse's Histoire de France, 1903). Influenced by the German geographers, he defined geography as a natural science, as a science of places and environments: "Geography has as its special mission to find out how the physical and biological laws that govern the world are combined and modified in being applied to different parts of the surface of the earth" (1913). He extolled the unity of this science, its "aptitude for not splitting up what nature has joined together," the descriptive method, a geography in which man is an active agent in the "combination" but differentially active according to his level of development and his cultural heritage (he introduced the concept of genres de vie). "The geographical character of a country is not something ordained in advance by nature . . . it is a product of man's activity, conferring unity on the materials, which by themselves do not have it."

Conscious of the relative and the contingent, Vidal de la Blache did not generalize--prematurely?--like his German colleagues. More than theories, he left behind models of analysis and description, difficult to imitate because of his remarkable qualities as a writer. The book of the historian Lucien Febvre, La Terre et l'évolution humaine, conceived before the First World War but only published in 1922, drew up an excellent balance sheet of that modern geography that arrived at maturity through diverse trends. Ridiculing the rigid determinism of certain Anglo-Saxon geographers, he credited Vidal with a "possibilist" doctrine.

This preponderance of two great names wrongly overshadows other artisans of geography, for example Elisée Reclus (1830-1905), a disciple of Ritter, who published from 1875 to 1894 the 19 volumes of a Géographie universelle[5]; the Russians P. P. Semenov (1827-1914), another disciple of Ritter, and A. I. Voeikov (1842-1916), one of the first climatologists (Les Climats du monde, 1884), whose Le Turkestan russe published in 1914 in Paris shows the direct influence of Vidal de la Blache; Lucien Gallois (1857-1941), whose work reflects the evolution of ideas, published in 1890 a thesis on the Géographes allemands de la Renaissance and in 1908 a remarkable study, Régions naturelles et noms de pays; Jovan Cvijić (1865-1927), a Yugoslav specialist in karst relief and author of a monumental work, La Péninsule balkanique (1918).

In England, Halford John Mackinder (1861-1947) began the teaching of geography in Oxford in 1887. The first edition of the celebrated Handbook of Commercial Geography of Chisholm dates from 1889. In Germany, Ferdinand von Richthofen (1833-1905) inaugurated the chair of geography in Leipzig (1883).

This crystallization of geographical science in Europe ought not make us neglect the influence of some other personalities, such as George Perkins Marsh (1801-1882), who, as a sideline to his diplomatic career, studied the role of man as an agent in the modification of nature and was the pioneer of resource conservation in the United States.

The Institutionalization of Geography

The history of geography then changes scale. It no longer traces individual attributes or geographers who became such after other backgrounds or teaching a geography that is tied to history. Geography acquired a much greater autonomy, and its history became institutional. It made its entry into the universities, and the professorships multiplied: at Oxford in 1887, Cambridge in 1888, in the German universities in 1874, and Rome in 1875. Institutes and laboratories of geography appeared, founded at Lille by Ardaillon in 1893 and Rennes by Emmanuel de Martonne in 1899. The Department of Geography of the University of Chicago was created in 1903. In 1896 there were 107 geographical societies, whose activities of publishing of reports, of encouragement of research, and of popularization were considerable. At the same time journals were multiplying (such as the Annales de géographie, Paris, 1891, and the Geographical Journal, London, 1893) and international institutions were created: the first Congress of Geography at Antwerp in 1871, and the Bibliographie géographique internationale in 1892. The International Geographical Union, founded in 1922, brought the geographers of the whole world together in congresses (eleven congresses since that of Cairo in 1925[6]) and tries to coordinate their activities in creating specialized commissions.

The 20th Century

The first decades of the 20th century were only the prolongation of that rather remarkable period that preceded them. Geography spread, and it took, in the majority of countries, a privileged position in primary and secondary teaching. It is the only discipline that provides an opening to the knowledge of the real world. Well before other disciplines such as sociology or demography, geography early found itself solidly established in the university, with all the consequences that result from that: research publications, doctoral theses, numerous jobs, diversification, and the appearance of "schools." Long confined to European countries or practiced by Whites, geography was internationalized. But apart from China and Japan, this extra-European geography was for a long time attached to European geography by a tie of a colonial nature. This situation existed not only in the colonies--French geography founded a family in French colonies, and English geography created the Indian, Australian, and Canadian geographies--but also in independent countries. Brazilian geography owes much to French geography; that is also the case with French Canada. The political and economic influences have changed and have been progressively diversified with time, and so certain independent countries thus receive influences from many geographical schools.

It is not possible to cite the names of all the geographers of this period who are leading toward contemporary geography. They have contributed to make geography a science of "multiple abodes." One no longer speaks of physical or human geography, still less of geography, but of geomorphology, climatology, geography of population, geography of the environment, and of rural, urban, economic, and historical geography. At a still grander scale, and more recently, have appeared geographies of capitalism, tourism, and disease.

This evolution is tied not only to the movement of ideas and to a trend toward specialization and application (the launching of applied geography dates from the 1950s), but also to the progress of documentation. The more exact and detailed maps, cadastral plans, and aerial photographs permit researches of exquisitely detailed cartography (such as the land utilization map of all the British Isles directed by L. Dudley Stamp during the 1930s). Population censuses, economic statistics, and surveys have multiplied the research topics and ideas.

At frequent intervals in the course of this growth, geographers have continued to ask themselves about geography and to pose some "eternal" questions:

1. On the nature of the relations between the earth and man. The geographer inevitably encounters the problem of determinism. Violently criticized either for philosophical reasons or because the deterministic causalities were improperly stated, the proposed substitutes, possibilism and probabilism, have not received a unanimous reception.

2. On the relations between geography and the other sciences, natural and social. The image of his discipline given to others by the geographer

is not always convincing. Certain awkward attitudes or assertions have often exposed geography to criticism or to incomprehension.

3. On the nature of geography. The debate, from the classical geographers to modern times, has never ceased to exist between an objective geography and a subjective geography, or between a "nomothetic" geography and an "idiographic" geography.[7]

The problems of regional division and of regional typologies have also held an important place--division into natural, geographical, agricultural, or socioeconomic regions--but without giving rise to new concepts or theories.

French geography, basking in the glory of Vidal de la Blache, has been one of the most flourishing geographical schools. Vidal's chair at the Sorbonne was divided into a chair of physical geography occupied from 1909 by Emmanuel de Martonne and a chair of human geography held by Albert Demangeon from 1912. In the same year (1912) Jean Brunhes took up a chair in the Collège de France. It was only in 1928 that André Cholley came to occupy beside them a chair of regional geography.

The disciples of Vidal de la Blache, and then those of his successors, all distinguished themselves with theses in regional geography: La Picardie of Demangeon, Les Paysans de la Normandie orientale of Jules Sion, and La Flandre of Raoul Blanchard. From 1906 to 1948 the last-named raised to a high place the renown of the Grenoble Institute of Alpine Geography.

The direct and indirect influences of the War of 1914-1918 and the creation of numerous posts led to a rapid development of French university geography. After the Second World War, regional theses ceased to be the most numerous with the multiplication of theses in geomorphology and rural geography and, more recently, urban geography (study of networks and of the relations between cities and their regions).

But this geography has remained marked by its past--both the recent and more distant past--which is shown by its trends:

1. A more than sentimental attachment to the unity of geography, which is shown by the requirements of the subjects of "complementary" theses, by the importance of solid works of systematic and regional geography by the same person, and by the rather large number of geographers changing orientation or specialization in the course of their careers.

2. A major orientation toward regional geography. Aside from theses, the Géographie universelle, published in 23 volumes from 1927 to 1946, symbolized that activity that permitted the publication of exemplary works: Les Iles britanniques by Demangeon, L'Europe centrale by Emmanuel de Martonne and L'Amérique septentrionale by Henri Baulig.

3. A privileged relationship with history, the methods and the historical spirit. Not only have history and geography always been associated in the cursus studiorum, but French geographers, like the majority of their European colleagues, have always thought that the importance of history was essential, if not determining, in the interpretation of geographical facts. Geography and history bring together

the names of Marc Bloch and Roger Dion in rural studies, and of Febvre and Demangeon (Le Rhin).

4. An orientation toward traditional societies and milieux, rural rather than urban and industrial, and toward the colonial milieux more than toward the very developed nations. With the passage of time, French geography appeared between the wars and just after the Second World War as having been rather isolated from foreign schools because of its orientations (colonial geography produced a constellation of great geographers: E. F. Gautier, A. Bernard, C. Robequain, J. Weulersse, and J. Richard-Molard), and doubtless also even because of its commanding position.

5. An attachment to the methods of literary description and coherent interpretation, fascinating in the extreme, more deductive than inductive. Especially between the wars, the geographer was supposed to have a beautiful style and to bring off beautiful descriptions.

6. A clear orientation toward the monographic genre. The monographs on regions, farms, and towns constituted the best work of geographers up to the 1940s. These monographs, valuable in themselves, have not often permitted the researches to be raised to the plane of generalizations and theory.

C. Trends in Geographical Thought

Detailed analyses of the works and of geographical publications make conspicuous the importance of comparable temperaments and of identical attitudes of thought and their permanence from one epoch to another.

Diversity of Currents

Geographers have put and are putting the accent on a geography that is a science of arrangement, location, and distribution of phenomena of whatever sort. Descended from the Kantian geography studying spatial properties, this current has persisted in German geography (e.g.--Alfred Hettner) and has strongly marked Anglo-Saxon geography and most particularly that of the United States. In that country geography is more and more inseparable from "regional science." It is there in close contact with economists and the theories and models of location.

The naturalist current is that of the physical geographers, geomorphologists and climatologists, then more recently oceanographers and biogeographers. They are rich in their methodological certainties, with a well-defined object of study. They have brought and are bringing to geography their scientific rigor, their exactness of vocabulary, and the experimental method. Their knowledge of natural realities leads them to give a certain weight to natural factors. Cvijic, Davis, D. W. Johnson, de Martonne, and Baulig illustrate this family.

From the middle of the 19th century, the socio-economic current put the accent on populations and societies rather than on environments or the relations between environments and societies. The centers of interest were or are the distribution of population, the differentiation of human groups, and the "diversity of the connections of production and the social connections that result from them in each geographical milieu" (Pierre George). This was formerly the geography of "genres de vie" and the human geography of Brunhes. Today it is a geography very close to demography, economics, and sociology. The Dutch geographical school, which calls human geography sociography, belongs to this current. Social geography, developed in West Germany (W. Hartke), is a branch of this kind of geography. If it gives a privileged place to social factors, it is no less concerned with analysis of landscapes.

Other geographers are attached to a current described as "landscapist," a current that is now expressed by the definition of geography as the science of the organization of space. They refer back to the Tableau of Vidal de la Blache, but they wish to be more scientific and oriented as much toward applications as toward the past. These geographers see in landscapes the geographical expression of societies and civilizations in relation to the natural environments. This trend originated in fruitful researches on rural (agrarian) landscapes. Maximilien Sorre, whose fundamental thought and work have suffered from not having had direct disciples, Belgian and Italian geography, and American cultural geography (Carl Ortwin Sauer) have represented or now represent this current.

Quantitative Methods and Literary Methods

The trends and temperaments are also manifested in the methodologies. Some people adopt with favor the measures and the indices that allow one to define objectively and to compare. They are also preoccupied with typologies and classifications, the basis of all science (e.g.-- classification of farmstead types by Demangeon). Since the 1950s a "new geography" founded on measurement and models has arisen in Sweden and has spread into the Anglo-Saxon countries.

Others, on the contrary, reject the quantitative apparatus. They insist that statistics have only relative value, and they justly declare that geographical reality is not reducible to numbers and that only literary description as fine as possible can give an account of it.

In 1899 Vidal de la Blache wrote: "Geography is an old science, but it periodically rejuvenates itself in proportion as it invigorates itself in live sources--that is to say, in the diversity of terrestrial scenes."

The evolution of geographical thought shows the importance of men and their temperaments. This evolution is made up of an alternation of periods of specializations and researches that seem antithetical and periods when geography sets fruit and rediscovers its own values.

At the present stage, the development of quantitative methods, satellite photographs, computers, and data banks open possibilities that were difficult to imagine only ten years ago in matters of universal instantaneous and evolutionary cartography, of comparative and correlative researches, and of elaboration of spatial models.

A new chapter in the history of geography is beginning.

Editor's Footnotes

(As an encyclopedia article, Pinchemel's essay did not have footnotes, but I have added a few where I thought they would be useful.)

1. The choice of this example is significant. During a lecture at UCLA in 1974, Professor Pinchemel said, in an aside, that he loves valleys; indeed, he married one. Mme Pinchemel was the former Geneviève Vallée. She also teaches geography in the University of Paris.

2. Cf. François de Dainville, Le Langage des géographes: termes, signes, couleurs des cartes anciennes, 1500-1800 (Paris: Editions A. et J. Picard & Cie., 1964).

3. André Meynier, Histoire de la pensée géographique en France (1872-1969) (Paris: Presses Universitaires de France, 1969), pp. 40-41, "La Pensée de Bergson."

4. Actually, Vidal published important geographical works in the 1870s and 1880s, as Professor Pinchemel noted in his paper, "Paul Vidal de la Blache," Geographisches Taschenbuch 1970/72 (Wiesbaden, 1972), 266-279.

5. The first weekly instalment of Reclus's Nouvelle géographie universelle was published in May 1875, and Volume 1 (all the instalments on Mediterranean Europe gathered together) is dated 1876.

6. Counting Antwerp (1871) as the first Congress, Cairo (1925) was the 11th, and the last, Tokyo (1980), was the 24th.

7. Cf. William R. Siddall, "Two Kinds of Geography," Economic Geography, vol. 37, no. 3 (July 1961), facing p. 189 (a note based on his Ph.D. thesis, "Idiographic and Nomothetic Geography: The Application of Some Ideas in the Philosophy of History and Science to Geographic Method," University of Washington, 1959).

Pinchemel's Bibliography (with emendations)

1. Richard J. Chorley, Antony J. Dunn, and Robert P. Beckinsale, The History of the Study of Landforms, or the Development of Geomorphology, Volume 1, Geomorphology before Davis (London: Methuen & Co., Ltd., 1964).

2. Paul Claval, Essai sur l'évolution de la géographie humaine (Cahiers de géographie de Besançon, 12; Annales littéraires de l'Université de Besançon, 67) (Paris: Les Belles Lettres, 1964).

3. François de Dainville, La Géographie des humanistes (Paris: Beauchesne et ses fils, éditeurs, 1940).

4. François de Dainville, Le Langage des géographes: termes, signes, couleurs des cartes anciennes, 1500-1800 (Paris: Editions A. et J. Picard & Cie., 1964).

5. Robert E. Dickinson, The Makers of Modern Geography (New York: Frederick A. Praeger, Publishers, 1969).

6. Lucien Febvre, La Terre et l'évolution humaine: Introduction géographique à l'histoire (Paris: La Renaissance du Livre, 1922).

7. Thomas Walter Freeman, A Hundred Years of Geography (Chicago: Aldine Publishing Company, 1961).

8. Thomas Walter Freeman, The Geographer's Craft (Manchester: Manchester University Press, 1967).

9. L'Information géographique, La Géographie française au milieu du XXe siècle (Paris: J.-B. Baillière et Fils, Editeurs, 1957).

10. Chauncy D. Harris, "Bibliographies and Reference Works for Research in Geography" (a mimeographed publication of the Department of Geography of the University of Chicago, 1967), recently published as Bibliography of Geography, Part 1, Introduction to General Aids (University of Chicago, Department of Geography, Research Paper 179, 1976).

11. Richard Hartshorne, The Nature of Geography: A Critical Survey of Current Thought in the Light of the Past (Lancaster, Pennsylvania: Association of American Geographers, 1939).

12. Preston E. James and Clarence F. Jones, editors, American Geography: Inventory and Prospect (Syracuse, New York: Syracuse University Press, 1954).

48

13. André Meynier, <u>Histoire de la pensée géographique en France</u> (1872-<u>1969</u>) (Collection SUP, "Le Géographe," ed. by Pierre George, 2) (Paris: Presses Universitaires de France, 1969).

14. John Kirtland Wright, <u>Human Nature in Geography</u>: <u>Fourteen Papers</u>, <u>1925-1965</u> (Cambridge, Massachusetts: Harvard University Press, 1966).

THE HISTORY OF THE METHODOLOGY
OF GEOGRAPHY AS A SCIENCE

Hermann Wagner

(This is a translation of Wagner's Lehrbuch
der Geographie, vol. 1 [Hannover: Hahnsche
Buchhandlung, 1920], pp. 17-25, "Geschichte
der Methodik der Geographie als Wissenschaft")

Antiquity.   To classical antiquity we owe a great wealth of
geographical information as well as the first attempts to collect and
assign this information to a particular scientific discipline. The word
Geography is derived from Greek origin; it does not occur before the time
of Aristotle.   In ancient times it referred only to the art of drawing
pictures of the earth's surface, and only gradually changed to its verbal
description.

Already in the time of ancient Greece and Rome the collectors,
advocates, and patrons of geography belonged to two distinctly separate
disciplines. This applied also to those few among them who were intent on
establishing and exploring a systematic field of study. The advocates of
the exact school represented by geometers and astronomers were concerned
with the shape and size of the earth, with the distribution of land and
water, the expanse of land masses, the location of countries or at least
the then known areas of them, and with the distance between them, which is
to say with the systematic organization of global areas. Their opponents
primarily took an interest in the influence of men and events upon the
various areas of the globe. Their actual field was history, and they
considered the knowledge of events and places a prerequisite to their
studies. Without mentioning the various predecessors I should like to name
Anaximander of Miletus as one of the earliest among the first of the above-
mentioned groups for the Ionian Period. He attempted to design the first
map during the first half of the 6th century B.C. Herodotus (c. 484-425
B.C.) is considered the father of historical regional geography and
ethnology. He dwelled in detail on the customs and historical events of
each particular country and lavished a great deal of attention on their
natural features. After the time of Alexander it was Eratosthenes (276-196
B.C.) of Alexandria who established the first systematic organization of
geography and thus provided it with a basic methodology. This methodology

determined scientific procedures for many centuries to come. In the first
books of his lost works he attempted to stimulate our powers of perception
and, hence, visualization of the shape and size of the earth. He then
sketched a map of the inhabited areas of the earth and introduced a grid
system which permitted the locating and reporting of positions. This was
an attempt to subdivide the areas of the globe into large geometric
sections and to describe them with regard to the countries and peoples
within. Although partially questioned by Strabo (68 B.C.-A.D. 24), a
Greek who lived during the reign of the Emperor Tiberius, the teachings of
Eratosthenes nevertheless form the basis of the famous seventeen books of
the Greek author. Strabo's work has been preserved almost in its entirety.
To call him the most important geographer of classical times, as has been
done during our time, means, however, to praise his importance particularly
for the historical branch of geography. Above all, and beyond his
methodological pursuits, he developed the idea that geography should be
considered a philosophy and thus (adopting the subdivisions of global areas
by natural boundaries as a guiding principle for geography) be entitled to
investigate the interdependencies of the natural features and the cultural
accomplishments of a given country as well as to demonstrate how these
achieved a certain perfection of customs and political institutions in one
and detained progress in another. Subsequent geographical descriptions
show little influence of Strabo's ideas. Periegesis, which means
consecutive description, as a writer would give of a coastline or highway
of a region as he travels along, was not conducive to the development of a
systematic geography. This style of description is represented by the De
situ orbis of Pomponius Mela (c. A.D. 40), whose geography books were
widely read during the Middle Ages and at the beginning of the modern era.
Consequently, it was the exact school of thought that emerged victorious
and not Strabo, and it is Claudius Ptolemaeus (c. A.D. 150) and not Strabo
who is considered the most important geographer of antiquity. His main
work, Guide to Geography, contrary to the tasks of chorography and
topography, sees the representative drawing of the earth's surface as the
essential function of geography. By making use of all the available
sources he offered a critical evaluation of the collected material for the
construction of maps of the inhabited areas of the earth as well as new
methods for a graphic representation of the earth.

The 16th Century and the Dissociation from Antiquity. We can dispense
with the Middle Ages. Graeco-Roman geography died. Neither during the so-
called patristic times that extended to approximately the year 1000 nor
during the scholastic period did the Middle Ages contribute anything of
lasting influence to methodology. The same can be said of Arab geography.
A general revival of the sciences brought about a rebirth of Ptolemy. His
geography, directly or indirectly, molded the field of geography until the
era of the great discoveries. Geography, or cosmography as it was called
from the time of the early Middle Ages, did not entirely rest in the hands
of astronomers. At that time the pursuit of humanistic studies was common
and did not exclude the pursuit of the mathematical sciences. The German
scientists of the 16th century excelled in these studies. Peter Apian's
widely read Cosmographicus liber (1524) is a typical example of the exact
approach to geography that bears a strong resemblance to Ptolemy's
interpretation of geography. It concentrated on the three principal

questions--orientation on the earth's surface (position-finding), calculation of the distance between two points, and mapping--but also took into consideration the regional geography of countries and the Abacus, which, like Ptolemy, listed towns and islands according to longitude and latitude and partly according to their political affiliation. About 1530, a German by the name of Sebastian Münster introduced a new type of representation as a cosmography and in the form of a comprehensive description of world history arranged according to territorial considerations. Münster directly followed the ideas of Strabo, although he did not fully realize all their implications. Now the different areas of the globe were being classified according to size, location, and other considerations. Assessments of productivity (fertility) and of the character and life-style of the inhabitants were being made. Methodological progress became clouded by deterioration by the time of Münster, with his partiality for regional history and the remarkable features of regions, peoples, and towns, which were no longer of any discernible relevance to the regional and local character and therefore to geography.

Cosmography comprised many different aspects. Not until the late 16th century did scientists pay due attention to the double meaning of the word "cosmos," which is universe (universum) as well as world order (ornamentum). The great Gerhard Mercator (1512-1594) attached an almost metaphysical quality to the latter and envisioned it as the ultimate goal of all scientific pursuit. He rendered a correct description of the earth's surface and thus made the most important contribution to the geography of that period. After 1600 the word "geography" established a firm hold over the various earth sciences, although within the systematic order it ranked below cosmography and above chorology and topography. A definite dissociation from the methods of antiquity took place, and with it the first distinction between "old" and "new geography" was made by the historians of geography. It was Philipp Clüver who initiated the new interpretation. What concerns us here is his insistence that the descriptions of countries should include reference to the existence of Man rather than his demands for a correct methodological description of the lands of antiquity. His Introductio in geographiam universam tam veteram quam novam was the focal point of all methodological considerations for a whole century. It must be said, however, that his geographical descriptions lacked a truly innovative approach.

Introduction of the Tripartite Division of Geography. By the middle of the 17th century a slim book of now classical content appeared and gave new directions to scientific research. This book was the Geographia generalis by the young Bernhard Varenius (1622-1650). It represents the foundation for a new approach to the methodological interpretation of geography that has recently been revived. It distinguishes a general geography, which deals with the total global picture and the various phenomena and formations according to certain categories in particular, from a special geography, which considers the sum of the various features of one particular place. The book essentially deals with the first of these two definitions. More important was the transition from the purely quantitative approach of Ptolemaic geography to a more qualitative and causative approach. What up to that time had been considered in the inadequate chapters on mountains or winds now became a separate branch of geography. For over a century the basic concepts of hydrography and meteorology and the first rough sketch of a morphology of the land surface of the earth were being developed. This means that Varenius not only founded the so-called physical geography but, by considering the affectiones terrestres as equal and next to the affectiones coelestes and the affectiones humanae, also suggested a new system of three subdivisions. The definitions remained vague, however, because Varenius still included the nautical sciences in general geography.

For a generation Varenius did not have a direct influence. Geographers continued to divide geography into two fields, to which they now attached the names of historical and mathematical geography. By the end of the 17th century and at first under various different names the three subdivisions were generally accepted. These subdivisions are still being used today. Of the early names the definitions of mathematical and astronomical geography have survived and with them their narrowly defined subject matter, which is the size and shape of the globe and the orientation on the earth's surface. Natural geography changed to physical geography by the middle of the 18th century. At that time historical geography considered the development of Man under three aspects--language, religion, and nationality. Of these only the last allowed a pictorial and, therefore, realistic representation. This must be considered the reason for the fact that the name of historical geography, which deals with Man's relations to the earth, was soon replaced by the designations of a "civil," a "historical-political," and finally a "political" geography.

The exact and the historical tendencies developed independently. With the exception of the Italian Joh. Bapt. Riccioli, who, in his voluminous Geographiae et Hydrographiae reformatae libri duodecim (Bologna 1661), which is essentially a comprehensive collection of material, demonstrated a definite inclination toward a methodological approach and therefore can be considered complementary to Varenius's work, few geographers engaged in the kind of general geography that the latter had proposed. And while during the time of Newton the science of physics underwent a spectacular development geography was barely touched by it. Again it was the astronomers and the mathematicians who focused on mathematical-physical geography, and, consequently, the purely physical aspects were not explored beyond what was known at the time. A comparison of the excellent Einleitung zur mathematischen und physikalischen Kenntnis der Erdkugel by the Dutchman Lulofs (1750) with the works of Varenius shows the obvious imbalances of the approach. During the middle of the 18th century a hot dispute ensued among the historical geographers as to whether, in matters concerning the classification of the earth, preference should be given to the natural or the political boundaries. This dispute indicated the necessity for a closer cooperation with physical geography. The Abriss der Geographie (1775) by the historian Joh. Chr. Gatterer attempted to unify the two separate schools. He did not consider these issues in depth, however, because adequate methods for a descriptive rendering of the natural features of the various continents as well as for the bulk of accumulated material had not been developed at the time. Historical geography joined the movement of the day and undertook to satisfy the growing interest in state and administrative institutions. Schlözer and Achenwall advanced the statistical approach, and, in connection with it, Anton Büsching achieved recognition with his statistical-geographical descriptions of various nations during the second half of the 18th century. In methodological terms this meant a growing eagerness among geographers to conduct exact research with the aid of any reliable methods and information, a tendency that in turn contributed to the accumulation of material foreign to geography. Büsching, who compared the natural features of the earth with those created by civilization, had not given sufficient consideration to the natural environment.

The Era of Humboldt and Ritter. Physical geography was greatly influenced by the valuable contributions of Linnaeus and Buffon, who had compiled and classified their observations of plants and animals. These first steps into the field of plant and animal geography were tentative and, for the time being, appeared in the form of annotations to the then typical form of description. The otherwise excellent descriptions of the Swede Torbern Bergmann and the lectures on physical geography by Immanuel Kant exemplified this situation. But the era of scientific exploration was in full swing, and under the guidance of men like de Saussure, Georg Forster, and Alexander von Humboldt the comparative investigation of the natural landscape forms and their global distribution was born. J. Fr. Blumenbach developed the first rough sketches of a systematic classification of the various human races according to geographical considerations (1775). E. A. Zimmermann presented the first animal geography (1778), and Humboldt wrote down his ideas concerning the geography of plants (1805). Far more important to geography, yet, was the fact that now the actual surface on which these organisms lived was being investigated. The face of the earth and her dimensional features were being given careful attention, and tools and techniques to measure heights and depths were being developed. The third dimension finally received due attention by the end of the 18th century, and we are greatly indebted to Humboldt (1769-1859)[1] and C. Ritter (1779-1859)[2] for having lent prime consideration to the study of the relief forms at a time when the skills of cartography were lagging far behind those of verbal description. Only therewith were adequate techniques for the proper description of the natural features of the earth found.

These two men are called the founders of scientific geography for having unified the two branches of geography, which, until then, had led a meaningless coexistence, and for having established relevant common grounds. Humboldt and Ritter envisioned a spatial classification of all living organisms and inanimate nature as the lofty goal for geography. Never before had such a program existed. Humboldt set forth to invoke a grand picture of the cosmic phenomena of the universe and presented a classification of plants according to their natural habitat but later abandoned the project—drained, as it seems, by the effort. Ritter, on the other hand, neglected numerous intermediate steps, and after having attempted a plan of the external forms of the land pursued those phenomena that lastingly influence a given locality, by subjecting them to a comparative investigation with regard to its population and over extended time periods. Yet while the revered geographer succeeded in doing justice to the two branches of geography, the investigation of inanimate objects as well as their impact upon mankind, which is to say, to the exact and the historical schools, he obviously did not realize how far he had digressed from his original theme, the earth, while he was immersed in the historical investigations of antiquity for his voluminous Allgemeinen vergleichenden Geographie[3] for forty years. He did not keep in touch with the numerous latest scientific achievements of physical geography—then in the hands of meteorologists and geologists—and, consequently, was not able to resolve the imbalance that he and his school, which he by preference had organized among historians, had created within geography. It had resulted in an excessive valuation of the historical aspects of geography. Some scientists went so far as to consider historical geography superior to general geography[4] and endeavored to assign the latter to the auxiliary ranks.

The Present. The above-mentioned stage of development lies behind us. The period of stagnation in scientific geography that followed the flowering under Ritter and Humboldt for a while drowned out that last era of exploration, which began around the middle of the 19th century and quickly revealed to us the still concealed portions of the earth's surface. With the wealth of new knowledge the natural sciences assumed leadership among the other disciplines. Not until then was geography accepted into the ranks of the academic sciences.[5] Under the influence of Oskar Peschel's works[6] the geographers of that period, most of whom had come from other disciplines, recognized the necessity of retrieving Humboldt's ideas and resuming the investigation of physical geography. Since Ritter's time our knowledge of the relief of the continents has undergone several changes. Cartography has made such progress that its productions are valued as independent objects of study and as sources for physical geography. Geology, it is agreed today, shares a common methodology with geography and is therefore considered a closely related field. Eduard Suess, Ferdinand von Richthofen, and others developed the basic concepts of geomorphology. Deep-sea exploration has opened up a new field of scientific investigation. The books and theories of Darwin created a bond between geography and the biological sciences. For several decades physical geography progressed more rapidly than historical geography, and soon voices could be heard that claimed geophysics as the main goal of geography.[7] These voices marked the culmination of a wave that had naturally arisen after a long period of neglect. Today's much greater number of scientific geographers has preserved us more than before from the continued pursuit of a one-sided course. With an expanded point of view the study of historical geography was resumed with accelerated interest. Influenced by the ideas of Friedrich Ratzel scientists have left the narrow area of European culture and are now investigating man in general as conditioned by his natural habitat.[8]

What Does This Historical Review Teach Us?

1. The general concept of geography of classical times has retained its validity. Yet aspects and points of view within the field have expanded and deepened.

2. From the beginning geography was a dualistic science, which means that the central inquiry concerns two fields, the nature of the earth and its influence upon mankind. Over the centuries each in turn benefited from special or exclusive attention while the other was temporarily neglected. Strabo and Ptolemy, Münster and Apian, Clüver and Mercator, Varenius and Busching, Ritter and Humboldt represented the general science of geography as pairs, which means that neither of them can be considered the sole representative of all aspects of the geography of their times.

3. Nevertheless, the general development aims at a bridging of this dualism, and it has become increasingly obvious that we are intent on considering the sum of all natural phenomena, organic and inorganic, as well as their dependence upon geographical factors and their interrelationships.

4. Considered from a purely methodological point of view geography in its beginning stages could not be considered a uniform science. During the last century, however, a tremendous accumulation of material created a complex field of many branches, and with it geography attained a certain degree of maturity. The continued task of the geographer remains to define and redefine the limits of the field and thus establish the territory of the auxiliary sciences that are striving for an independence of their own.

5. Marginal disciplines are crowding in on geography. Certain fields of investigation overlap and impinge upon its scientific territory. This situation creates the danger of dilution and corruption of geography by the auxiliary disciplines. Consequently, the geographer constantly finds himself forced to reconsider his methodology of investigations regarding the local distribution of all natural phenomena, their natural features as well as their historical implications. Münster, as we know, as well as Büsching and Carl Ritter, digressed from geography to history and Varenius, Bergmann, and Kant in other directions.

6. Like the other natural sciences geography has entered a stage of development in which scientists are attempting to solve the questions of existing phenomena by investigating how they evolved. Considerable attention is being given to the element of time. Ritter resorted to history for an understanding of the ever-present influences of geographical factors. Today we no longer regard the short timespan of human tradition as the "geographical present," beyond which the geographer is not allowed to look. The close cooperation with geology, the investigation of periodic imbalances and secular changes of various phenomena and their influences--climate, for example--show that the same requirement is asserted in other branches of geography.[9]

Wagner's Footnotes (with emendations)

[Editor's note--I have included only the last nine footnotes in this
section. These concern modern geography from Humboldt and Ritter
onward. The earlier footnotes (thirty in number) concern the history
of geography from classical times through the 18th century, and the
references would be relatively inaccessible and unimportant to the
modern student.]

1. Carl Christian Bruhns, Alexander von Humboldt, eine wissenschaftliche
    Biographie, 3 volumes (Leipzig: F. A. Brockhaus, 1872). In the third
    volume Humboldt's influence on the various branches of geography is
    shown.

2. Friedrich Marthe, "Was bedeutet Carl Ritter für die Geographie,"
    Zeitschrift der Gesellschaft für Erdkunde zu Berlin, vol. 14 (1879),
    374-400; expanded version published in 1880 (Berlin: D. Reimer).
    See also Geographisches Jahrbuch, vol. 8 (1880), 530. The biography
    of Ritter by Gustav Kramer--Carl Ritter, ein Lebensbild, 2 vols.
    (Halle: Verlag der Buchhandlung des Waisenhauses, 1864-1870; 2nd ed.,
    1875)--gives few facts for making a judgment of his scientific
    importance.

3. Carl Ritter, Die Erdkunde im Verhältniss zur Natur und zur Geschichte
    des Menschen, oder Allgemeine, vergleichende Geographie, als sichere
    Grundlage des Studiums und Unterrichts in physikalischen und
    historischen Wissenschaften, 2 vols. (Berlin: G. Reimer, 1817-1818).
    Second edition, greatly expanded, 19 vols. (in 21 parts) (Berlin: G.
    Reimer, 1822-1859). Vol. 1 (1822) on Africa; vols. 2-19 (1832-1859)
    on Asia (incomplete).

4. See Hermann Guthe's introduction to the first three editions of his
    Lehrbuch der Geographie (1868-1874). After Guthe's death in 1874,
    Hermann Wagner edited the subsequent editions (4th ed., 1879), and so
    this was the forerunner of Wagner's Lehrbuch. "Geography," Guthe
    begins, "teaches us to know the earth as the dwelling-place of Man."
    "Historical geography is the true geography."

5. For details see Hermann Wagner's "Bericht über die Entwicklung der
    Methodik der Geographie" (title varies), Geographisches Jahrbuch,
    vols. 7-14 (1879-1890/91). Ferdinand von Richthofen's inaugural
    address, Aufgaben und Methoden der heutigen Geographie (Leipzig:
    Verlag von Veit & Comp., 1883), has turned out to be especially full
    of methodological significance.

6. There comes to mind especially Peschel's book, <u>Neue Probleme der</u> <u>vergleichenden Erdkunde als Versuch einer Morphologie der</u> <u>Erdoberfläche</u> (Leipzig: Duncker und Humblot, 1870).

7. Georg Gerland's introduction to <u>Beiträge zur Geophysik</u>, vol. 1 (1887), and Hermann Wagner in <u>Geographisches Jahrbuch</u>, vol. 12 (1888), 418-444 ("Georg Gerlands Methodik der Erdkunde," pp. 418-444 of Wagner's "Bericht über die Entwickelung der Methodik und des Studiums der Erdkunde (1885-1888)").

8. See Hermann Wagner's "Bericht über die Entwickelung der Methodik und des Studiums der Erdkunde," <u>Geographisches Jahrbuch</u>, vol. 14 (1890-1891), 371-462. Since that time a detailed synopsis of the newer methodological views has not appeared.

9. Another fundamental interpretation of the progress of development in the various branches of geography in the course of a century was presented by Alfred Hettner in his inaugural address at Tübingen, "Die Entwickelung der Geographie im 19. Jahrhundert," <u>Geographische Zeitschrift</u>, vol. 4 (1898), 305-320, that went back to even earlier times. However, repeated examinations allow me to retain the above-mentioned views of these developments.

CONTEMPORARY GEOGRAPHY

Alfred Hettner

(This is a translation of Hettner's <u>Die</u>
<u>Geographie</u>: <u>Ihre</u> <u>Geschichte</u>, <u>Ihr</u> <u>Wesen</u> <u>und</u>
<u>Ihre</u> <u>Methoden</u> [Breslau: Ferdinand Hirt, 1927]
pp. 90-109, "Die Geographie der Gegenwart.")

Contemporary geography differs considerably from the geography taught by Alexander von Humboldt and Carl Ritter and by their pupils and disciples. A complete break with the old school of geography, however, did not take place. When the two founders of a more advanced geography died in 1859 a whole generation disappeared with them. Darwin's book, <u>Origin</u> <u>of</u> <u>Species</u>, which was to have a tremendous impact not only on biology but also on the geography of Man and the geography of plants and animals, was published in 1859. The actual shift of the general concept of geography did not occur until ten years later. It was initiated by Peschel's <u>Neue</u> <u>Probleme</u> <u>der</u> <u>vergleichende</u> <u>Erdkunde</u>.

The acquisition and colonization of foreign territories by various nations had become increasingly successful. New means of transportation had been developed and greatly contributed to the comfort of traveling. The advancing weapons technology created a steadily growing advantage over the primitive races. Canned goods, new knowledge in the fields of medicine and hygiene, as well as many other accomplishments eased travel conditions in general and facilitated advances into foreign and sometimes hostile territories. This was the time when the Antarctic and the Arctic circles, when Central Asia and the interior of the African and of the remaining continents were explored, when knowledge was gained about the areas that, until then, had remained blank spots on the map. By that time most of the explorers had some scientific training, and their journeys can be considered research expeditions. These expeditions were faced with fewer external difficulties than the earlier ones and usually limited their investigations to special areas and specific problems. They began to specialize in either botany, zoology, or geology, in anthropology or archeology. A new type of traveler appeared, the man who traveled either for knowledge or for pleasure. If he happened to be a knowledgeable person or happened to be a scientist himself he would often bring home valuable discoveries. World travel made easier by circumstance became fashionable, and globe-trotters, not always agreeable human beings, could be seen crowding hotels and the steamships headed abroad. A special class of tourists were the alpinists, who were no longer content with climbing the unconquered heights of our Alps but were scaling the mountain ranges outside of Europe and thus providing new knowledge of alpine regions.

The large-scale maritime expeditions, usually accompanied by scientists, have become rare. The Austrian Novara expedition of sixty years ago deserves mention. The marine expeditions designed for deep-sea exploration were of a different nature: in the 1870s the English Challenger expedition, the American Tuscarora, the German Gazelle, followed by the German Valdivia expedition under the guidance of the zoologist Chun (1898-1909), and, recently, the German Meteor expedition, which is limited to the southern Atlantic Ocean but has crossed and probed that region several times already.

Successful advances toward the Far North and the Far South were undertaken, and finally the Poles were reached. The first important step toward Arctic exploration was taken when Adolf Erik Nordenskiöld circumnavigated Asia in the Vega in 1878-1879. He found the long-sought Northeast Passage but had to realize that it was not suitable for regular maritime traffic. This was feasible only up to the mouth of the Ob and possibly the Yenisey. The northern coastline of North America had been established in the 1850s by excursions from both sides into the Banks Island region. At that time the Americans Kane, Hayes, and Hall and the Englishman Nares contributed to the further exploration of the Canadian Arctic Archipelago. In 1902 Sverdrup managed to penetrate its northern expanse and reach the west, and between 1903 and 1906 Amundsen finally achieved the North West Passage after he had skirted the continental coast. Knowledge of the east coast of Greenland was gained by a German polar expedition under Koldewey and by a Dane named Koch. Peary solved the mystery of the northern coast by sled and thus verified the insular character of Greenland. In 1883 A.-E. Nordenskiöld advanced towards the icefields of inner Greenland, and in 1888 Nansen achieved the first crossing, which later was to be followed by others. Already in 1873 an Austrian expedition under the guidance of Payer and Weyprecht had discovered Franz Joseph Land, and between 1893 and 1896, having come from the New Siberian Islands and headed west, Nansen crossed the Arctic Ocean on the Fram and on foot and eventually reached 86° N. In 1900 the Italian Cagni, a companion of the Duke of Abruzzi, managed to move somewhat beyond this point. On the 6th of April 1909, after a long march across the frozen ocean, Peary reached the actual North Pole or at least a point in the immediate vicinity.

Antarctic exploration had lagged since the journeys undertaken by Ross. G. Neumayer and Clements Markham finally kindled a new interest in Germany and England, respectively. This interest derived its most immediate impulse from the voyage of a Norwegian whaling ship. A young naturalist by the name of Borchgrevink had accompanied the ship and was the first to set foot on the Antarctic continent. In 1898-1899 Gerlache and Arctowski in the Belgica were the first to winter over in the South Polar ice, and in 1899-1900 Borchgrevink advanced on dog sleds to 78°50' S. During the years that followed, expeditions of various nations competed for knowledge of the South Polar region: a German expedition under E. von Drygalski, English under Scott, Scottish under Bruce, Swedish under Otto Nordenskjöld, and a French expedition under Charcot. The English expedition fared best insofar as it was able to advance via the Ross Sea, where the ocean penetrates most deeply toward the pole. By sled Scott managed to travel across the icefields of the Ross Sea and move into the mountain range south of it. Other expeditions followed. After a long journey through the highlands Shackleton managed to reach 88°23' in 1908, while his companion Davis located the South Magnetic Pole. Amundsen finally reached the South Pole in 1911, and Scott got there a few weeks later. Shackleton's efforts to cross the Antarctic continent in 1914 were doomed from the start.

It cannot be denied that in the discovery of the North Pole as well as the South Pole sporting ambitions prevailed over scientific interests. The scientific yield, however, was considerable, and although sizable areas of the various regions remained obscure the gross characteristics of the distribution of land and water in the Arctic and Antarctic were established. We know today that few islands exist north of Northern Asia, the Canadian Arctic Archipelago, and Greenland. Most of the area is covered by water. Contrary to the initial assumption this ocean is covered by ice for most of the year. Antarctica, on the other hand, does not just consist of various islands but is a continent that ranges in size somewhere between the continents of Australia and South America. Terra Australis, which had played such an important role in the history of geography, finally achieved a modest resurrection. The weather stations adopted by the civilized nations in 1882-1883 became increasingly significant to the scientific investigation of the polar climate and natural conditions generally.

Now let us consider the interior of the various continents. Europe has been carefully mapped and has become an object of detailed and specialized exploration. Only the climbing of the Alpine peaks continues to offer true satisfaction to the explorer.

Considerable progress has been achieved as far as our knowledge of Asia is concerned. Siberia is now in the stage of detailed investigation, which is no longer the concern of German but of Russian scientists. The completion of the Siberian railway allows travelers to cross the continent and gives them at least a rough impression of the country. Since the 1870s the region of Turan, the lowlands surrounding the Caspian and Aral seas and their border regions, has been included in the Russian domain and has thus become available for scientific exploration. While during the 1860s the Hungarian orientalist Vambery could only travel there disguised as a Muslim pilgrim, no more than a Russian passport is required nowadays to enter the region by railroad. Beside the explorations of Mushketov and other Russians, foreign travelers have added to our knowledge. Von Radde, in particular, has deepened our knowledge of the lands of the Caucasus.

Old Turkey became more accessible to travel. Asia Minor was largely explored by visitors with archeological interests, and Heinrich Kiepert made maps from their drafts. Philippson thoroughly investigated the western regions. The War contributed to our knowledge as well. Information about Syria and Mesopotamia was gained in similar fashion. Arabia, however, partly because of her natural conditions and partly because of the intolerance of her inhabitants, has resisted the European intruders; even today travels to the interior are travels of exploration in the true sense of the word. The map still contains large blank areas. Iran occupies a position somewhat between those of the above-mentioned countries. Travel remains difficult and precise data are not yet available, but a large number of visitors from various nations have fairly clearly elucidated the character of the country.

Around 1860 Central Asia was still largely unknown. A. von Schlagintweit had been murdered in Kashgar in 1857. Then several expeditions began to move in simultaneously from the north, the south, and finally the east. Needless to say, Russia and England, who owned the adjacent territories, supplied most of the scientists. England was able to enlist the help of the Indian Pundits to explore Tibet, which was generally closed to Europeans. Most of the credit, however, goes to Sven Hedin, who explored the Taklamakan Desert in the Tarim Basin and the interior of Tibet. Also to be mentioned are Semenov as the first explorer of the Tien Shan and Przheval'skii as the first explorer of the Gobi.

By adopting European civilization Japan took a decisive step toward scientific exploration. Once a preliminary survey, taken by E. Naumann and other European scientists, many of whom were Germans, had secured a basis for the various fields of scientific investigation, the Japanese increasingly took over the exploration of their own country. The ease of travel and the charm of the land and its inhabitants began to attract many European and American travelers. These tourists were able to gain a general, if superficial, impression of Japan.

The opening up of China advanced much more slowly. When F. von Richthofen on his journey (1868-1872) supplied the foundation of our scientific knowledge of the region he was confronted with severe difficulties. Recently several main routes have become accessible by train, and it may be assumed that European scientific investigators have gained access to large regions of the country. The Chinese have refrained from joining these expeditions.

Systematic survey and a network of observation stations were established in the British colony of India. Naturally their distribution is less dense than that in Europe. It resembles our facilities before about a hundred years ago; the scientific methods, however, are excellent. Yet much remains to be done in the field of scientific investigation, which is never completed with the mere gathering of official data. The interior of Southeast Asia has been accessible to scientific investigation only since the 1860s. Surveys and explorations have been started by the English in the British West and by the French in the French East. Far behind is the exploration of Siam. Thanks to the efforts of the Dutch government considerable progress has been made in the East Indies. Java, for instance, is as well known as most of the European countries. Scientists, particularly botanists and zoologists, are facing a new and promising realm for investigations. As far as modern scientific exploration is concerned the Philippines remained untouched under Spanish rule. The advent of American rule brought significant changes.

Of New Guinea and Melanesia little was known at the beginning of our era. The very nature of these areas and particularly of Melanesia represented severe problems to scientific investigation. With the onset of European civilization rapid advances were made, and Germany in particular may proudly claim knowledge of her former colonies.

The settlement of the Australian continent advanced from the eastern shores westward across the mountains and covered the arable land. A second wave of settlers composed mostly of prospectors in search of gold moved into the interior from the southwest. Topographic and scientific information increased accordingly. In 1872 a telegraph cable was strung from the Gulf of Carpentaria to the south coast. Soon afterwards, the continent was successfully traversed from east to west and vice versa.

As far as New Zealand was concerned, the situation was similar. Hochstetter, in particular, contributed to our knowledge of the country. At the same time the smaller islands scattered throughout the Pacific Ocean were explored.

On the African continent exploration was in full swing. Only a few expeditions can be cited here. Rohlfs crossed the Sahara on several extensive journeys (1865-1879), Nachtigal moved from the Sahara into the Sudan (1869-1874), Schweinfurth explored the region of the Bahr-el-Ghazal and the Uelle. In East Africa Livingstone traveled from Zanzibar to Lake Nyasa in 1866 and from there into the interior, where he disappeared without a trace. Stanley, a reporter for an American newspaper, tracked him down in 1871. Stanley was now inspired, and in 1874 he crossed Bagamojo and started on a long journey that was to take him to the Congo via Lake Tanganyika and from there to the Atlantic Ocean. He had accomplished the crossing of the African continent. Already a little earlier (1873-1875) an Englishman named Cameron had managed to cross Africa further south. These two journeys essentially completed the discovery of the Congo region in its principal features, and the last important hydrographic puzzle had been solved. In the years that followed, Savorgnan de Brazza, Pogge, and Wissmann, among others, explored the regions of the southern tributaries of the Congo. Wissmann was the first to traverse these latitudes from the West (1880). Further south the Portuguese Serpa Pinto had crossed the continent from Benguela to Durban (1877-1879). Stanley undertook another important journey into the forests of the northern Congo region and from there to the east coast (1888-1889).

By then the exploration of the African continent had undergone a change. Stanley's crossing of Africa had induced King Leopold of Belgium to create the State of Congo. Approximately at the same time Germany decided to acquire several territories (Togo, Cameroun, Southwest and East Africa) as colonies, the English and the French had expanded their colonial territories, and the Portuguese were bent on establishing firm control over their holdings. The political division now influenced the nature of geographical exploration. Each nation concentrated on the exploration of the territories within the boundaries of her colonies, and the cultivation and pacification brought about by European rule over large parts of Africa generally aided and improved the conditions for travel and exploration. Officers and civil servants made route-maps on their travels, meteorological observation stations were installed, and scientists began to pursue detailed and specialized investigations.

In North America the connection between East and West was established during the middle years of the last century. Yet the West largely remained unexplored. Eventually it became important to the federal government to collect at least the basic information on the unknown territories. Under the guidance of men like Hayden, King, Powell, Wheeler, and Whitney several expeditions moved west. Above all the West promised minerals. Therefore, geological and topographic investigations were fortunately combined. In 1879 the Geological Survey was established, in which men like Gilbert and Dutton worked under Powell's leadership. These expeditions contributed not only to the knowledge of these particular regions but to the universal store of knowledge in tectonics and geomorphology. Research along similar lines was instituted in Canada somewhat later.

For some time Middle and South America trailed behind North America. Journeys of a true exploratory nature were the river trips through the interior like those of Crevaux and the exploration of the Xingú by boat by von den Steinen and his companions. Here, too, detailed and specialized investigations began to replace the large scientific expeditions, particularly in the fields of botany, zoology, geology, and geography. But few regions had actually been surveyed. Many scientists, young and old, especially Germans, worked on these natural history investigations. Out of a very large number of names connected with travels in the northern Andes, I should like to mention only the volcanic studies of Reiss and Stübel.

During the first half of the century information about the configuration of the earth had been obtained through a series of geodetic measurements, and now the geodetic work became still more standardized. In 1861 Baeyer established the central European geodetic survey, which became the European survey in 1867, and in 1886 it was widened to the international survey (under the direction of Helmert). Along with it systematic gravimetric measurements were adopted. These led to the discovery that the earth is not a true sphere but an irregular figure, a geoid. The deviation from true sphericity is, however, smaller than originally assumed.

The determination of terrestrial positions continued to be obtained from astronomical sighting. It eventually greatly profited from telegraphy, which permitted the transmission of accurate time signals and rendered the longitude measurements more precise. The geodetic surveys were followed up with careful, systematic mapping. Maps of ever larger scales were produced, in Germany at 1:25000, and more recently maps have been made at the scale of 1:10000 or even 1:5000 for the use of cadastral surveys. Beside these official surveys, private photogrammetric surveys were made of single mountains. Several colonies instituted land surveys, although at smaller scales. It must be remembered, nevertheless, that numerous maps of large regions of the globe were compiled from simple route maps with only a few accurate positional data. They still contain many errors. Recently some surveying has been done from airplanes. This method promises good results, especially for inaccessible woodlands.

Great advances were made in the theory and practice of map projection, especially with Auguste Tissot's Mémoire sur la représentation des surfaces et les projections des cartes géographiques (1881), which was introduced into Germany by Hammer. Hermann Berghaus and Vogel perfected the technique of reducing maps of large scale to the smaller scales of atlases and outline maps, and a method of producing maps from route sketches was developed by Heinrich Kiepert, August Petermann, and Hassenstein. The art of terrain representation and large-scale general survey maps reached a certain perfection at that time. The development of improved reproduction techniques, copper engraving as well as the less pleasing but more economical lithograph and the intermediate stage of copperplate printing proved very important.

Physics and chemistry added new dimensions to our concept of the earth's interior, especially with the theory of the critical phase change and the discovery of radium and helium. To geography they were only of indirect importance and mattered only insofar as they influenced the structure of the crust. The study of volcanoes, earthquakes, glaciers, rivers, and lakes was increasingly guided by the methods of physics and, as a result, was partly incorporated into the field of geophysics. Most of the research, however, continued to be carried out by the methods of geology and geography.

Geology was able to utilize the development of two auxiliary sciences: microscopic petrography and paleontology. The latter had been stimulated by the theory of evolution. Petrography and paleontology greatly improved the stratigraphic dating of rock layers; until then the formation and composition of rock material had been neglected. It is only recently that scientists have paid more attention to them and have been able to gain a more precise idea of the history of the development of the earth and of paleontology, which is the science of the nature of geographical configuration of past stages of development.

Tectonics, the science of the inner structure of the earth's crust, is part of geography. In the middle of the 19th century it was only in its initial stages of development. The decipherment of the small simple mountains was well done, but the interpretation of the large mountains was still under the influence of the rising eruptive mass. Two American scientists, Dana and LeConte, were the first to see the formation of mountain ranges as a process in which the rock material remains passive. They attributed the effects to the contraction of the surface as a consequence of the cooling of the earth. Eduard Suess expanded this theory for the Alps (1875) and eventually, in the form of an excellent survey, for the entire globe (Das Antlitz der Erde, 5 vols., 1888-1909). Almost simultaneously with the publication of Suess's 1875 work, Albert Heim, developing the ideas of his teacher Escher, presented his book on the folding of one particular section of the Alps (Mechanismus der Gebirgsbildung, 1878). These interpretations did not entirely withstand the scrutiny of advancing science. The theory of folding was expanded to the theory of nappe. Suess's assumption that all vertical movements of the crust produce depressions has proved one-sided if not incorrect; the currently prevailing assumption is that rising indicates block movements. Beside the block movements in areas of geological faulting, large bucklings are now being investigated. Once again scientists are attributing considerable importance to the magma. Geotectonics has become a field of detailed and specialized research, which not only has to consider the stratification of rocks but also their age and composition; this means that tectonics is branching away from geography. It also means that geography will have to be content with accepting the results offered by tectonics.

Along with tectonics the morphology of the land surface developed. It was long neglected by geology and found its development in geography, although the latter can come to no clear understanding of forms if it only describes them and does not understand their causes. English geologists had done a certain amount of research, but a clear understanding of the cause and effects of the formation of valleys, the basis for our understanding of larger formations, did not exist. It was still assumed that valleys represented fissures caused by the rising of mountains. It was Rütimeyer, an anatomist from Basel, who demonstrated that the large Alpine valleys had been created by erosion. Especially productive for geomorphology were the investigations in the American cordilleras where the barrenness of the land and the full view of the layering of the rocks favored research. Powell presented the explanation of transverse valleys. Gilbert worked out the first theory of erosion. Credit for further geomorphological investigations goes to von Richthofen, who advanced the study of the topography of arid regions. Research on the features of glacial topography followed. It was Torell, a Swede, who recognized that the deposits on the lowlands of northern Germany had been caused by the Scandinavian ice sheet. Detailed investigations ensued; Penck thoroughly explored the glaciation of the German Alps. It had originally been assumed that surface form depended solely upon the nature of the prevailing rock structure. Now climate, a research field of geography, was recognized as an important factor as well. As research advanced the impact of surface erosion and transformation not only on the valleys but on the total landscape became obvious. By means of a deductive method the American Davis tried to formulate a system for the erosion of surface features and gained considerable attention. His as well as Peschel's attempts to solve morphological problems by a comparative study of maps have proved unsuitable.

Pedology, the study of the material composition of the soils of the different regions of the world, had been badly neglected by geographers. It was Richthofen who corrected the situation. While Western European research on soils was chiefly connected with the underlying rock, a Russian scholar [Glinka] and a German-American named Hilgard recognized the dependence on climate.

Hydrology was carried on in several fields. The methods of physics and geology were being applied to the study of firn and glaciers. The study of groundwater supplies and springs was carried on with the aid of geological surveys. The laws of hydrodynamics were being investigated by hydrologists. The study of lakes, or limnology, was carried on separately and greatly profited from the investigations of Forel, a scientist from Geneva. This fragmentation of efforts could not satisfy the keen interest that geographers had in hydrology. A comprehensive hydrographic survey of the earth did not yet exist.

Great progress has been made in the field of oceanography, especially in the field of deep-sea exploration, which became possible when Brooke invented the deep-sea lead in 1854. Not only did the various expeditions of the 19th century contribute to our information but also the routing of cables across the ocean floor. Our hazy and rather fanciful conceptions of the bottom of the sea were replaced by a basically sound although still only general understanding. The depth investigations included temperature measurements and various physical observations. The study of the surface of the ocean advanced simultaneously. Until recently ocean currents had been determined by the difference between the log and the actual position as determined by astronomical observation. Now the observations of water temperature, salinity, and density supplied us with additional information. Yet oceanography still is considered a subsidiary science in the larger field of the general earth sciences and not a special branch of geography. The interpretation of the ocean as a geographical entity has seen only a few attempts, such as Schott's study of the Atlantic Ocean.

Climatology was supplied with much new material through the progressive settlement and europeanization of the world. New stations were established, and the longer the series of observations the more reliable became the averages. Here, too, changes of concept slowly took place. At the beginning of the period meteorology had undergone a change insofar as the synoptic weather maps introduced by Buys Ballot in 1854 had become widely known and were accepted as a basis for our understanding of weather conditions and weather forecasting. Climatology no longer consisted of general investigations of atmospheric conditions but was able to assess the actual present conditions: it moved from climate to weather. Consequently climatology and meteorology moved apart; climatology remained within the field of geography, and meteorology became a separate science based on physics yet intent on branching away from it because of the immensely complicated nature of atmospheric conditions. Now prime consideration was given to the dynamic aspects of research and less to statistics, and today we speak of meteorology as the natural history of weather. The objectives of research also changed at that time. The key to understanding was to be found in the distribution of atmospheric pressure rather than in the winds, because it is atmospheric pressure that influences winds, precipitation, and temperatures. Recently this assumption has been challenged by the so-called theory of the polar front. This theory is favored by many and indicates a trend toward returning to Dove's theory. Naturally these conceptual changes affected climatology to some degree. Yet climatology has remained a statistical science that works with averages and relies too heavily on weather station observations.

The geography of plants and animals was moving ahead in two directions. One was defined by the theory of evolution proposed by Lamarck at the beginning of the century. It had been given little attention at the time. Now Darwin and especially Wallace were applying it to animal geography as did Hooker and Engler to the geography of plants. The fauna and flora of islands and mountains in particular were being investigated in their genetic or genealogical aspects. The occurrence not only of similar but of related species in separate areas seems to indicate an original relationship, and it may be assumed that the related species developed differently because of spatial separation. Not only do the former land connections support these interpretations but climatic conditions do so as well. The occurrence of similar species at the cool heights of separate mountains suggests that during the Ice Age a similar climate prevailed in the lowlands between them. Hypotheses abounded, especially among the botanists and zoologists, many of whom did not hesitate to let continents disappear for the love of a single plant or animal species. Eventually more disciplined methods prevailed. Genuine geological evidence had to be presented to prove the assumed existing conditions of past periods.

The other direction pursued a more fundamental approach, with what Häckel called ecology. It belonged chiefly to plant geography. Humboldt in his Ideen zu einer Physiognomik der Gewächse [1806] had described the various forms of vegetation and vegetation formations along with the various families of the plant world. Others followed. Grisebach in his beautiful book, Die Vegetation der Erde (1872), had investigated the influences of climate on vegetation. The progress of plant physiology enabled Schimper and other botanists to understand the dependence of all plants on their natural environment from a botanical point of view. They first investigated the Arctic region and the tropics and finally regions of a less pronounced climate. Certain misconceptions developed that are now being challenged.

For a long time the science of animal geography remained reluctant to consider these ecological aspects. So far only certain ocean regions have been studied with special considerations given to their particular influence upon the local animal population. Geographers in particular were painfully aware of the lack of ecological information. With his Tiergeographie auf ökologischer Grundlage [1924; English edition by W. C. Allee and K. P. Schmidt, Ecological Animal Geography, published 1937] Richard Hesse eventually responded to an obvious demand.

The geography of Man began in this period with Peschel's critical review of Ritter and the Ritter school. His criticism was not altogether fair because Peschel had applied an incorrect comparative methodology and had not given sufficient consideration to the development of Man, which Ritter clearly emphasized. Peschel disputed Man's obvious dependence on nature. His critical evaluation did cause a shift of concept, however, and changed the essentially teleological interpretation to a causative one, and thus opened the field to detailed and specialized research. The progress of physical geography supported the geographical interpretation of Man, for only a precise knowledge of the typical features of a given object will allow us to assess its dependence on another. For example, the influence of coastal formations on traffic conditions and settlements was not clearly understood until we learned to distinguish among the various types of coastline. Regional geographical research now includes the geographical conditions of Mankind. Already in the 1870s Kirchhoff had given consideration to them in his geography lectures. The first volume of Richthofen's China (1877) contains impressive descriptions of various centers of civilization and the migration of people in Central Asia. Consequently one cannot speak of the founding of the geography of Man by Ratzel. The first volume of Ratzel's Anthropogeographie (1882), originally intended as a single-volume work, represents an investigation of the influence of the natural environment upon Man and thus was merely an introduction to the geography of Man. Not until the second volume (1891) were the distribution of Mankind, population conditions, and settlements subjected to a searching examination. In his anthropogeographical investigations of ethnology Ratzel overemphasized the impact of communications on the distribution of peoples and underestimated the importance of the natural environment, which Bastian had rather overrated. Here and there attempts were made to exclude the geography of Man from the science of general geography as an alien element or else to confine it to those facts that had visibly affected the land. These efforts did not prevail, however; on the contrary, anthropogeographical research began to flourish after the aspects of physical geography had created a sound base for further investigations. Today research predominantly engages in anthropogeography rather than physical geography. It has to be said, however, that few—possibly too few—scientists consider the larger historical context and tend to focus on the facts and figures of colonization, population distribution, settlements, traffic, economy, and, recently, on the problems of political geography. The loosened ties to history have not been strengthened. The reason for this may be the continued refusal of historians to consider the importance of geographical conditions in history rather than a tendency among geographers to underestimate historical development.

Already during the first half of the 19th century the various earth sciences and many travel descriptions had enhanced our knowledge of the nature of the earth and of various countries in particular. This knowledge increased steadily. For a long time geographers refrained from making use of it. Ritter and his school had neglected the study of the natural environment. This situation was untenable. The great discoveries in the polar regions and in the interior of the continents excited the interest of the geographers. Several geographical societies were founded. New geographical journals appeared on the market. Others, like Ausland, shifted their emphasis to geography. It became evident that the science of geography would be absorbing a steadily growing wealth of material; it also became increasingly obvious that current geography, which had been geared to history and statistics, was at a loss to handle the information on those countries that had had neither a recognizable history nor represented recognizable states. It was high time to engage in a more comprehensive study of the general nature of countries. For several decades geography had concerned itself solely with the collection and recording of new material. Toward the end of the 1860s the need for a more comprehensive scientific understanding surfaced and prevailed.

In 1866 Oskar Peschel's Neue Probleme der vergleichenden Erdkunde first appeared in Ausland as a series of articles and soon was published separately as a book (1870), at about the same time as Elisée Reclus's La Terre [1868-1869]. Peschel's spirited if somewhat superficial book caused a tremendous response and inaugurated a new era for the science of geography.

In his <u>Geschichte der Erdkunde</u> Peschel expanded the concepts of geography beyond those of Ritter. In his <u>Neue Probleme</u> he attempted to explain the formation of fjords, islands, deltas, and other natural features on the surface of the earth, as well as the distribution of the steppes and deserts, by a comparative study of maps. He began to solve problems that Ritter and his school had never considered. It remains to be said, however, that the introductory polemics aimed at Ritter are unfair and represent a purely literary feud over the word "comparative." Most of the so-called new problems had been tackled before by geologists and other scientists. The value that was being attached to the comparative study of maps warranted a certain concern. Most of the geographers of that period were not familiar with the literature of the natural sciences, especially not with foreign publications, and therefore did not share this concern. They were excited by the promises of a new field of science and by the apparent accuracy and simplicity of the investigations as well as by the elegance of the definitions; only now geography seemed to assume the character of a true science. It had roused itself from the rigidity in which it had existed and began to bring forth buds and blossoms. The ideas of Humboldt and other great explorers were finally being accepted. Science entered geography and especially into regional geography. The anthropogeographical approach to the study of Man was revitalized. A time of eager scientific exploration and successes ensued. Since Ritter's death geography had been excluded from the universities. Now Peschel was appointed professor of geography in the University of Leipzig, and soon other Prussian universities and the University of Strassburg followed the example. The universities of several other German states and those of several other civilized nations established departments of geography at about the same time.

These efforts to reinterpret geography moved somewhat beyond their targets. A true spirit of conquest prevailed. The attractiveness of the new problems, the scope of which could not yet be defined, as well as incorrect interpretations of the concepts of methodology guided by the very name of science and by conceptual considerations rather than by the actual scientific developments tempted them to pursue fields that converged with the fields of other sciences. Geography, it was occasionally maintained, should not occupy a place next to the various specialized sciences but a place above them, like philosophy, and, also like philosophy, should be considered a comprehensive empirical science. It is not surprising that the related sciences objected to these demands and that geologists and meteorologists did not care to see their sciences reduced to subdivisions of geography. The geographers themselves recognized the danger of uncontrolled expansion and, consequently, of deterioration. A desire to remove anything alien, to limit the research to a certain field, and to define the methods of research and presentation eventually replaced the zeal for exploration.

A few members of the older generation of geographers and even more so the historians who were about to lose geography as a serving maid preached a return to Ritter and meant thereby to give a refined treatment to Mankind and its history. This was not tenable. The nature of the earth's surface and its different development in different countries or landscapes not only demand a descriptive but a scientific and genetic approach that the systematic sciences cannot supply. This applies to anthropogeography as well. It cannot be successfully established in poor soil. Should the uninhabited regions with no history be excluded from geography? Not one among today's geographers would care to support this conclusion.

The other proposition advocated that geography should concern itself solely with the study of nature and exclude the study of man entirely and that it should limit itself to indicate only a rough design for historians, economists, and anthropologists. This position stands in complete disagreement with the history of our science, in which Man had always occupied an important position--a more important one than the natural environment, in fact. Man as well as nature are undeniable components of the general features of a country. Few regions of the earth remain untouched by Man. Most of them carry his imprint. For this reason alone the study of Man must not be neglected.

Geography now adopted Richthofen's ideas as guiding principles. To be sure, the thoughts that he expressed in the conclusion of the first volume of his work on China were very one-sided on the relation of geography to geology, and even in his inaugural address at Leipzig that launched modern geography he did not succeed in offering a sharp methodological definition of his approach. As his research showed, he was not always able to incorporate his own methods successfully and later occasionally lapsed into the old patterns of general geography. However, science, above all else, depends upon the basic concept and to a lesser degree upon sharp definitions. The basic concepts had been established. Geography renounced the claim to a holistic interpretation of the earth and no longer demanded the incorporation of other sciences like astronomy, geodesy, geophysics, geology, meteorology, and others. These had been independent for some time. Geography, finally in tune with its historical development, embarked on the exploration of the diverse features of the earth's surface, with a thorough investigation of continents, countries, regions, and localities. It did not limit itself to a purely mathematical or ethnological-historical approach of the various schools of classical times but made use of the rich knowledge of nature gained from the development of the natural sciences and the information brought home by the various scientific explorers. It excluded those natural phenomena that do not show an immediate connection with the nature of a particular country as well as ethnological, historical, and statistical facts; the object of investigation was the character of a country, landscape, or region, its manifestation in their particular natural environment, and the conditions for Man. At the same time and by comprehensive comparative methods it began to explore the earth's surface to search for an understanding of its diverse formations and the diverse character of its many regions in the larger context of the earth as a whole. Geography was no longer only a specialized study but had become a comparative and comprehensive science as well. A few of the older scientists clung to what they themselves termed a more dualistic conception of geography; it favored two separate problem areas, the exploration of the earth as a body of nature and as Man's habitat. However, the conviction that scientists should not pursue two tasks of a basically different nature but take an integrated approach to the field spread fast. Resorting to an archaic expression Richthofen called geography a chorological science. The development might be interpreted as a return to Ritter. If so, it certainly was not a return to the narrow approach we encounter in his late works and in those of his pupils but to his original propositions. Modern geography still follows his original program but includes the modifications brought about by recently gained understanding and modern technical accomplishments.

Since the geography of peoples and political geography were admitted to the fringe regions of geography and after a time of relative calm, disputes have arisen over the question of whether human geography should content itself with the investigation of topographic and physical evidence or whether it should include cultural evidence insofar as it was conditioned by the natural surroundings as well. The study of landscapes as formulated with great fanfare by Passarge simply attaches a new name and does not contribute new insights to the field. The same applies to his "comparative study of landscapes," which had been formulated before. Banse's "new geography," on the other hand, would represent a genuine departure from old thought patterns. He considers scientific geography to be only a first step and that the proper condition of geography is rather as an art, but his ideas have generally been discarded by most geographers.

The reforms spread from Germany to other countries. France, in particular, accepted the new ideas. Although Elisée Reclus had published a general physical geography simultaneously with Peschel's Neue Probleme, he remained a true pupil of Ritter; his great regional geographical work, Nouvelle géographie universelle, shows the spirit of Ritter. It was Vidal de la Blache who became the leader of the new wave. This new direction comprises a general physical geography as well as anthropogeography with the emphasis placed on the study of landscapes and the natural environment, for which the French, due to their particular aesthetic inclinations and their excellent gift for presentation, have a natural talent. Several geographical monographs of high quality have been done on France and the French landscape.

England has been slow in accepting the new methods of geography. Although eager travelers and outdoorsmen and, therefore, it would seem, naturally disposed to engage in the study of geography, the English seem to lack what is essential to the study of geography: sophisticated and proper scientific work habits, as they were developed by other scientists, and the spirit of synthesis that is a prerequisite to geography. Herbertson did import the new movement from Germany and did establish a School of Geography at Oxford; yet the Royal Geographical Society remained cool to his efforts. Only recently has the need for a new approach to the science of geography been voiced.

The United States of America, under the leadership of Davis, concentrated on geomorphology, so that the geography professorships are often linked with geology. Only recently has a certain tendency toward innovation been observed.

Changes are taking place everywhere. Several countries are engaging in specialized studies with general geography confined to the old methods. Others are actually engaged in the process of reform. It would take too long to dwell on all of them here.

Today geography is undergoing vigorous development. The beginning of the 19th century showed similar symptoms and was followed by decline because of the narrow scope of the reforms. There is no need to anticipate a similar deterioration today, because the present development is much stronger and more general and has even taken over the universities, thereby making the possibility of a richer, more scientific development easier. Geography as a science has taken firm roots, from which fresh knowledge will filter into our schools and daily life. However, our goals remain far from our accomplishments. It is time to reassess, not to be detained by success.

Note

Hettner did not employ footnotes, but the interested reader can find most of the relevant materials in the notes appended to the other papers.

GEOGRAPHY AND TRAVEL IN THE 19th CENTURY:
PROLEGOMENA TO A GENERAL HISTORY OF TRAVEL

Hanno Beck

(This is a translation of "Geographie und
Reisen im 19. Jahrhundert: Prolegomena zu
einer allgemeinen Geschichte der Reisen,"
Petermanns Geographische Mitteilungen, vol.
101, no. 1 [First Quarter 1957 , 1-14.] Beck's
paper was dedicated to the memory of
Heinrich Schmitthenner.)

In the 19th century travel came to be considered as a tool of
geography.[1] The methodological limitation of geographical research to the
earth's surface and a relative geographical conception of space were
accepted. Geography could not rest until she thoroughly knew her objects,
and thus most journeys of discovery simultaneously became scientific
expeditions. If someone wanted to look exclusively at journeys of
discovery, he would have to limit himself and exclude, for example, the
insights of the more literary-oriented travel literature.

The pure history of discovery can hardly cover all questions of
interest to geography.[2] Carl Ritter characterized this by a utilitarian
and logical guiding motive by treating the history of discovery as an
aspect of enlarging our knowledge of space. In this article it is
attempted to consider the history of discoveries from the point of view of
geographical science. Thus it is not important to mention a lot of names.
Instead, it is important to prove that travel and discovery journeys were
increasingly scientifically oriented. This is proven by the fact that
within the period of the history of science travel was increasingly
influenced by geography. Similar to Ritter's concept of the enlargement of
our spatial knowledge, we can prove a development that we consider as a
guiding motive, i.e., the growing influence of geography on the development
of travel within the periods of the history of scientific geography. Thus
our task is to demonstrate how this influence is manifested. One necessary
consequence of this view is to enlarge the history of discoveries to a
history of travel in general,[3] where the rather neutral word "travel"
stands for journeys of discovery, scientific expeditions, and other
journeys. Thus this essay also attempts a summary of the journeys within
the periods of the history of the discipline in order to demonstrate their
dependence on geography.

The 19th century saw the end of "exploratory geography." Yet the tendency toward it had been very important in the past. Until the 19th century the reports of Greco-Roman antiquity were valued as sources. Barth, for example, still took "his Herodotus" along on his travels, and even Ritter and Humboldt tried to gain insights from the oldest sources, as the European Middle Ages had done little to extend the rather important developments of ancient geography. The interest that Ritter, Humboldt, Mannert, and Ukert, as well as Rennell and Malte-Brun showed for ancient geography is a clear demonstration of the necessary, source-critical evaluation of classical geographical reports, which could at first only theoretically and then also practically be confirmed or surpassed through the development of their own geographical exploratory plans. Thus the foundations of classical German geography cannot be understood if one fails to consider the preparatory work of the 18th century.

1. Pre-Classical German Geography (1750-1799) as the

First Step Leading to the Developments of the 19th Century

The 18th-century attitude toward geography is acknowledged in literature with remarkably vague statements. Nevertheless, during this period the foundations for the future development were being built. Many important questions were already solved by J. M. Franz, Büsching, Gatterer, the two Forsters, and, last but not least, by Kant. But the results were not absorbed into the general consciousness. Classical German geography often dealt with problems that were considered new although in fact the previous work was simply unknown. Demarest was the first to explain the formation of valleys through erosion, and he made the important distinction between high and low plateaus. But in Germany, for example, he was only known to R. E. Raspe, and only Humboldt may have known more about his importance than his other German contemporaries. Lomonosov researched the chernozem soils, considered volcanoes as local phenomena, pointed to convective streams of the atmosphere, and developed an "aerodynamic machine" to facilitate research of the highest air strata. He recognized cartographic analogies, promoted the idea of the Northeast Passage, and anticipated the actualist principle of geology. But there was a lack of continuous scientific research by professors and students at the universities. J. M. Franz, after Barthel Stein the second German professor of geography, was already an excellent methodologist, but he did not train any geographers because at the time there were no professional opportunities for geographers. Thus his idea to have geographers in government service in the territories was important, as it would secure a place in the universities for geography as a science.

From the point of view of the history of science Göttingen University, which was founded only in 1737, was the center of the most significant geographers: Franz, Tobias Mayer, Lowitz, Büsching, Gatterer, G. Forster, and Heeren.[4] The popularity of educational travel permitted university professors to teach Apodemics, the art of travel. Tobias Mayer influenced Karsten Niebuhr. Some rulers sponsored actual expeditions. August the Strong sent Hebenstreit and Ludwig to Tunisia and Algeria in 1732-1733.

Charles de Brosses, president of the Dijon parliament, pursued the history of travel and required that expeditions be guided by scientific knowledge. He inspired Bougainville, who took the botanist Commerson and the astronomers Verron and Zeichner along on his trip around the world (1766-1768). Other plans by Bougainville failed, especially a trip to the Arctic during the time of the French Revolution. However, this first French world traveler recognized already that many geographers of his time were playing with theories or betraying reality to systems, and he expressively coined the phrase that geography should be a "science of data or facts."

The geographer Alexander Dalrymple (1737-1808) summarized at the end of each volume of his <u>Historical</u> <u>Collection</u> <u>of</u> <u>the</u> <u>Several</u> <u>Voyages</u> <u>and</u> <u>Discoveries</u> <u>in</u> <u>the</u> <u>Pacific</u> <u>Ocean</u> (2 vols., 1770) the geographical desiderata from the travelers of his time in order to inspire them to new discoveries. Among other things he collected information from the South Seas for the East India Company in Madras and also corresponded with Charles de Brosses. When the American Revolution threatened England's mercantile position, he drew attention to the imaginary southern continent. He was proposed as a participant in a big South Sea expedition by the Royal Society but was refused by the Admiralty when he demanded to lead the expedition. Afterwards, James Cook (1728-1779) was commissioned to lead this project, and in 1768-1771, 1772-1775, and 1776-1779 he led remarkably modern expeditions, which were, with regard to their preparation, execution, and evaluation, the first great scientific expeditions.[5] The two Forsters, who accompanied Cook on the second voyage around the world, incorporated the field of travel research into the development of geography; they concentrated on landscapes and fascinated a generation, just as Pallas did with his Siberian journeys that also inspired Humboldt.

On June 9, 1788 the Association for Promoting the Discovery of the Interior Parts of Africa was founded in London, and from then on facilitated planned expeditions. The considerable success of this Association is partly due to Sir Joseph Banks, a participant in Cook's first voyage, but above all since 1790 to the first significant English geographer, James Rennell (1742-1830), who had considerable experience as a field researcher from his service as a cartographer for the East India Company. His maps of India and Africa belonged to the few thorough works of their kind. Like D'Anville, Rennell reinterpreted the African maps anew and equipped travelers with clear orders and good maps that were specially designed for this purpose. He tried to evaluate the often meager reports of the unsuccessful travelers of the African Association and to investigate their factual content. The first travelers for the Association, the American John Ledyard and the Englishmen Lucas and Houghton, did not reach their destinations. Ledyard died, while Lucas, who had lived as a slave in Africa, did bring back some important news. Already in 1790 Major Houghton was influenced by Rennell. Mungo Park, the most successful traveler of the

Association, explored the Niger on his journey in 1795-1797. With his expertise he combined luck. He returned home, and already in 1799 his notable travel account appeared in German. Rennell calculated Park's route and was able to considerably improve his general map of Africa.

In 1795 Johann Friedrich Blumenbach (1752-1840) arranged for the first German, his pupil Friedrich Hornemann, to be admitted to the Association. The Association granted him a scholarship for a new field of study, primarily Arabic, in Göttingen and obtained permission for him to cross France, which had been at war with England since 1793. Hornemann was the first to succeed in crossing the Sahara, but he died in the process, so that only part of his results could be published.[6]

Blumenbach, who personified the polyhistorical tradition right into the 19th century, inspired and sponsored travelers; Langsdorff, Hornemann, Horner, Collmann, W. L. von Eschwege, Prince Max zu Wied, G. H. Roentgen, Seetzen, J. L. Burckhardt, Bialloblotzky, and Humboldt were his pupils or were inspired by him.[7] It speaks well of Blumenbach that he also proposed Johann August Zeune, who was one of the most important geographers of the transition period between the 18th and the 19th centuries, for an expedition.[8]

Numerous expeditions--the journeys of Shaw, Bruce, and Peysonnel, the Geneva Center of Natural Research, Saussure's glacial research, and the often mocked Büsching, who nevertheless proved able to describe foreign countries in an interesting fashion--all these belong to the colorful panorama of the 18th century, which can by no means be dismissed with such concepts as rationalism or utilitarianism.

At the end of the 18th century there appeared in Germany several geographical journals that could not be matched by any foreign publications. The Germans gave extensive consideration to foreign articles, and translations of foreign books appeared frequently and remarkably fast. They also considered the precise research of some Frenchmen, whereas in France, for example, German scientific aspirations and accomplishments were known insufficiently and the research of Englishmen was followed only in a few instances.

Following the advice of Franz von Zach (1754-1832) the duke Ernst II had an expensive planetarium built on the Seeberg near Gotha, which he equipped with the most modern instruments from London. Zach corresponded with astronomers and geographers in both Europe and America. He edited his Monatliche Correspondenz zur Beförderung der Erd- und Himmelskunde, which thenceforth critically evaluated the results of scientific expeditions. Many travelers, among others U. J. Seetzen and A. von Humboldt, received from von Zach and his successors Bernhard von Lindenau, Encke, and Hansen the necessary astronomical knowledge, which then guaranteed the quality of their research. It is to the efforts of von Zach that contemporary geography is obligated for much inspiration and for a scientific as well as popular magazine, which found wide attention.[9]

Nevertheless the verdict of the following period of Romanticism on this age prevents many researchers from keeping an unbiased stand. Thus, for example, at that time there developed the beginnings of a history of geography.[10] But one should be careful of generalizations[11]; the century is much more problematic and full of ideas. Just as Vico preceded historiography, so Demarest in France, Lomonosov in Russia, Bergmann in Sweden, and the two Forsters in Germany were precursors of geography. What is remarkable is not the lack of geographical observations, morphological research, etc., but the lack of a systematic synopsis of the geographical material in view of the generally existing encyclopedic efforts. To have attempted such condensation, then, which philosophically comes from the 18th century, is the merit of Ritter and Humboldt, the classical German geographers.

2. Classical German Geography (1799-1859)

Classical German geography--the Erdkunde (in the sense of geography) of Ritter and Humboldt--grew out of the previous work of the 18th century. Influenced by their generational position, both had decisive and determining impressions of the years between 1780 and 1800 that remained influential up to their last years. Thus they lived through the confrontation of geography and statistics and observed how the "pure geography" gave a new direction to the science. Ritter had already sketched before Zeune in 1806 the idea of natural regions.[12] But Zeune is nevertheless an important figure as the most important geographer in Berlin before Ritter and Humboldt. Through his geographical lectures since 1810 he prepared the way for Ritter's work as a lecturer in geography. Ritter's career is hard to follow as he began by following the more general traditions of the past. He started by recognizing the importance of pure geography, learned to avoid its mistakes, and separated geography and history, which in itself is of great merit. He too wanted to travel, but his career obligations and the narrowmindedness of the Prussian authorities prevented larger journeys. He methodologically considered the concept of space in geography, recognized the necessity of logical organization, and, like Humboldt, recognized the importance of travel for the furtherance of methodology.

Humboldt, on the other hand, started with the most modern geographical teachings of the 18th century--i.e., with scientific expeditions. He gratefully recognized Georg Forster as his teacher but distinguished himself scientifically from his teacher. The appreciative observation of the environment that Ritter took over from Salzmann and Pestalozzi as a pedagogic principle was widened by Humboldt just as by Ritter. But Humboldt avoided Ritter's deviations, the study of which is rather interesting. Humboldt developed in a straight line. He started by following Forster, saw and described landscapes, and in 1796 had the stroke of genius of his "idea of a physics of the earth." He smoothly and methodically separated history and geography, probably just like Ritter by following Kant. Humboldt's conception of geography is effective and is distinctly recognizable and by no means drowned in natural scientific generalizations. He recognized many geographical problems but did not always write them up. His Kosmos is above all the work of a geographer, only here the physical description of the earth is enlarged to a physical description of the entire globe. Humboldt's great American journey (1799-1804) is a truly decisive event as well as a shining example for all subsequent travelers. The thoroughness of its scientific preparation, execution, and evaluation remained a solitary achievement that was often imitated but never reached.[13]

All the geographical developments of the time were overshadowed by Ritter and Humboldt. This became rather clear after 1827 when Humboldt remained in Berlin and thus in the same place as Ritter. Through his lectures in the Singakademie [the largest concert hall in Berlin] and the university in 1827-1828, Humboldt demonstrated that he could not only think methodically but also teach methodically, just like Ritter. But probably he would never have enjoyed being a university professor, and he always tried to avoid such offers. More important is Ritter's university position. As an excellent teacher and scientist, he inspired an entire generation of geographers with his knowledge and oratorical talent, and thus he became the most successful teacher in the annals of geography.[14]

Thus, for example, A. P. Efremov audited Ritter's lectures between 1840 and 1842 and then started his career as the first scientifically trained professor of geography at Moscow University in 1844 with a course of lectures in "general geography." Unfortunately, he had to retire in 1847 because of his connection with the Slavophiles [a group of Russian intellectuals who resisted excessive westernization]. Thus, between 1847 and 1855 there were no more geographical lectures in Moscow. More decisive were the activities of Ritter's most important Russian pupil, P. P. Semenov of the Imperial Geographical Society of Saint Petersburg, of which he wrote a three-volume history. With him Russia had someone to inspire and act as an example as an explorer ("Semenov-Tian-Shanskii") and as an organizer. The expeditions of Przheval'skii, Potanin, Grumm-Grzhimailos, and many others were facilitated or supported by him.

Ritter and Humboldt were conversant with the great problems of geography in their time and selflessly helped young travelers. In Germany there was a lack of funds and political influence that often supported the foreign travelers. But Germany had, through Ritter and Humboldt, the most highly developed geographical research. "England should furnish the money, Germany the men to advise and act," Ritter once wrote when he tried to organize an expedition.[15] German travelers were wanted as they had no secondary political aims but only showed dedication to scientific aims. They did not want "to establish quarters for the Prussian army," as a Frenchman acknowledged ungrudgingly; they wanted to work scientifically.

Reciprocity developed between geography and travel research. Lichtenstein's reports from South Africa became important for Zeune as well as Ritter. In order to write a regional description of Asia Minor, Ritter worked relentlessly to make the area more accessible. Kiepert traveled in Asia Minor himself and with tireless industry marked every enlargement of topographic knowledge on his masterful maps, and Humboldt, up to a very advanced age, registered the progress of younger travelers in order to give greater precision to his morphographic concepts and to improve his view of nature.

Unfortunately the work of the African Association slackened after Hornemann's death. Rennell withdrew in 1802, although he still followed the progress of exploration. After Henry Nicholl failed in 1805, G. H. Roentgen, who was recommended by Blumenbach, travelled in 1809. He was supported by an English lady, and wanted to push from Mogador to Timbuktu, but he was murdered in 1811 by fellow travelers in the province of Haha. Thus the Association could not solve the Niger problem, which was then first resolved by the German geographer Reichard from Lobenstein (in the manner of Humboldt) on the grounds of logical considerations. For political reasons the English government interfered with the developments. At first they did not establish their own organization or society, but they wanted to know the commercial and political value of the unknown African territories in order to make political decisions. Thus modern European colonialism resulted from this research, a development that German researchers recognized relatively late. As secretary for colonial affairs, Lord Hobart inspired in 1803 the first expedition financed by the English government. Mungo Park travelled in 1805-1806 to solve the question of the Niger mouth and the possibilities of trade, and Rennell advised him before the expedition. The German astronomer Oltmann deserved much credit for the interpretation of the astronomically based results when he proved to the French geographer Walckenaer that Bowdich's corrections of Park's results were not justified. Oltmann already supposed that Rennell did not evaluate the results as they contradicted his cartographic system. Thus, among others, Rennell had not considered the correct position of Murzuk by Hornemann, which was later confirmed by Vogel and von Beurmann. In 1822 the next great expedition commenced with Denham, Clapperton, and Oudney. In 1825 Denham and Clapperton brought the first geographical news from the heart of Africa. In the same year Clapperton and Lander were sent out anew. They went from the Bight of Benin to Bussa and Kano and wanted to press on to Timbuktu, which was entered in 1826 by Alexander Gordon Laing. Lander completed the research of Clapperton, who died in 1827. In 1830 Lander travelled the then unknown portion of the Niger between Bussa and the mouth.

The year 1830 is a decisive turning point in the exploration of Africa. The French started the foundation of their second colonial empire in Algeria. In 1836-1838 Moritz Wagner pushed through to Algeria while it was still being conquered by the French, and he published an excellent travel account. The English suspiciously watched the French taking possession and now gave a scientific frame to their own political interests by founding the Royal Geographical Society. In 1833 the property and experience of the older African Association were taken over by the new society, which also merged with the Raleigh Club (1827-1854), whose members were explorers and sailors. After 1849 German geography worked closely with the Royal Geographical Society. Christian Karl Josias von Bunsen established the ties as Prussian ambassador in London, and as a close friend of Humboldt he expressed Humboldt's wishes, which found much attention, as did all mentions of German exploration geography. Sir William Hooker and Edward Sabine, who were often consulted by the African Association, were connected with Humboldt. Soon another German beside Bunsen represented his fellow countrymen in the investigation of the earth: August Petermann.

Ritter's pupils mainly explored Asia and Africa. Humboldt knew them all. Their letters were exchanged by the two geographers. In the New World many referred to Humboldt's name. But we cannot separate Ritter's from Humboldt's influence, and again and again we recognize their cooperation. Alexander von Humboldt, a geographer, developed modern travel techniques in an exemplary fashion, so that a Frenchman said of him that he discovered the art of travel. In the methodological unity of his work Carl Ritter tried to include all results and thus pointed out to travelers the gaps in the knowledge of the earth.

The geography of this time also advanced the cartographic and pictorial representation of foreign countries.[16] Technical aids--for example, the Nollet apparatus--were hardly known. Thus travelers were obliged to become landscape artists. Thus the description of landscapes in both words and pictures became highly developed. Sketching belonged to the necessary education in that time, and Humboldt and Ritter learned it. Almost all travelers brought back sketches, which almost always could be only partly published in the works themselves. Frequently their realistic sketches were redrawn at home by retouchers and thus falsified to some degree.[17] Many sketches of both Humboldt and Ritter are known. Humboldt considered it worthwhile to collect these sketches. Later on he encouraged training in the use of daguerrotypes. It is little known that photographic equipment was used very early on scientific expeditions. The expedition artists were soon accompanied by photographers. They did some excellent work that still draws attention and is much admired today.

In France physical geography was further advanced and enlarged by Malte-Brun to a science of regions in obvious dependence on Ritter and Humboldt, but also in considerable independence.[18] As a literary figure, as a political firebrand, as a poet and as an excellent stylist this Danish-born understood the problems of exploration geography. In 1807 he founded the Annales des voyages, de la géographie et de l'histoire, which appeared until 1871, and, as a skilful negotiator, he removed many difficulties in the preparation of the founding of the Paris Geographical Society (1821).[19] This society, with Humboldt as a member of the central committee, also inspired many enterprises and some prizes, although it lacked the means of the Royal Geographical Society. On the advice of Malte-Brun the Society edited the "Receuil des questions adressées aux voyageurs et à toutes les personnes qui s'intéressent aux progrès de la géographie," which was justifiably called a "negative encyclopedia," as it made an effort to emphasize to travelers the gaps in geographical knowledge. These publications were one of the most important attempts to direct the travel of the period according to scientific geographical principles. The trip of Dumont d'Urville, who was well known to Humboldt, between 1825 and 1829, was, for example, well prepared by his friend Arago.

Also significant is the competition for participation in the Austrian "Novara" Expedition (1857-1859). Carl Scherzer, Moritz Wagner's fellow traveler in the United States and Middle America, was one of the first who was suggested.[20] The minister of finance, Bruck, met Scherzer when he was working for the Vienna Academy and then recommended his participation. Through his friend Moritz Wagner got to know of the proposed expedition. King Maximilian II of Bavaria, a pupil of Carl Ritter, interceded and recommended Wagner to the Austrian king, who refused this personal recommendation. Scherzer said later with some justification that through the participation of Wagner the Novara Expedition could have become the German counterpart of Darwin's "Beagle" Expedition (1831-1836). Surely Wagner was rather unpopular in Vienna for the relentless description of Austria's political weakness in his books, but Humboldt's decision was decisive. Humboldt counseled and worked out an outline for the journey.[21] Humboldt suggested R. Avé-Lallemant, who had asked for his personal recommendation, as a participant.[22] It can be documented that he did not support Wagner for the Novara Expedition because he had more important plans for him. In 1856 or 1857 Wagner visited Humboldt and Ritter in Berlin and got detailed instructions of his tasks for a new trip to Middle and South America. Maximilian II paid Wagner for the trip on the recommendation of a commission, one of whose members was Justus Liebig, who in his youth had been a protégé of Humboldt and who thus fulfilled Humboldt's wishes. This expedition made Moritz Wagner the best expert on Middle America up to the time of Karl Sapper.

The Russian world travels, which should be considered closely, were scientifically well prepared. Georg Forster was supposed to participate on the first trip led by Mulovskoi, which unfortunately failed because of the war with Sweden (1788-1790). The economic and political aim of the Russian enterprises was the exploration of the sea route to the possessions of the Russian American Company and the establishment of trade with Japan and China. Mulovskoi's pupil, J. F. Krusenstern, in particular, clearly sketched this in a memorandum and then was entrusted with the first Russian cruise around the world (1803-1806). Participants were Blumenbach's pupils Horner, Langsdorff, and Collmann, as well as G. H. Tilesius from Leipzig, one of the most interesting personalities in the history of travel, and, last but not least, F. F. Bellingshausen and O. von Kotzebue, who became famous through their own journeys. Blumenbach, whose activities in this period should not be underestimated,[23] sent intructions to G. H. von Langsdorff, for which he, as so often, had been specifically asked. The study of these travel instructions is particularly informative and shows that Blumenbach had genuine geographical interests.

With two trips around the world, 1815-1817 (with Chamisso and Eschscholtz) and 1823-1826 (with Eschscholtz, Lenz, Hoffmann, Preuss, and Siewel) Otto von Kotzebue followed Krusenstern's successes. Adelbert von Chamisso, who later was one of the founders of the Berlin Geographical Society (1828), discovered on his trip of 1815-1817 the generational changes of the Salpidea [invertebrate marine animals], without however drawing any scientific conclusions. His travel account has a pronounced geographical flavor and was considered for a long time as a model without equal.[24] During Kotzebue's second trip the physicist Emil Lenz used for the first time the bathometer that he had invented. And on the fourth Russian cruise around the world, under F. F. Lutke 1826-1828, some scientists (e.g., F. H. Kittlitz) articipated who gave it lasting importance.

There were travelers who consciously started from geographical considerations. This is shown by the collaboration of Richthofen, Semenov and the Schlagintweit brothers with Ritter and Humboldt.[25] These young scientists wanted to work geographically when they designed their ambitious plan for the exploration of Inner Asia. Semenov was a pupil of Ritter and later translated the Asia volumes of his Erdkunde into Russian and partly enlarged them to a considerable extent in the process. He also enjoyed the confidence of Humboldt, who suspected volcanoes in the Tien Shan and in 1856 suggested a visit to Vesuvius. At that time Humboldt told the prospective traveler that he could only die in peace after he had received some rock samples from the Tien Shan.[26] The Schlagintweits are indebted to Humboldt's sponsorship for their journey; through Bunsen he opened up India for them and motivated the Prussian king to grant financial aid.

It is of the greatest importance that many of the scientific travelers found their way into universities, for example, R. von Schlagintweit, who taught in Giessen, but also Barth and Moritz Wagner, and later Ratzel and Richthofen. Carl Ritter's most important pupils, the Frenchman E. Reclus and the Russian P. P. Semenov, dominated scientific geography in their respective countries after excellent accomplishments in field research.

The age of great discoveries had avoided the technically most difficult to reach areas. The more important is thus the improvement of maps between the years 1799 and 1859. The fundamental scientific books of Humboldt and Ritter elucidated the main directions of contemporary travel research: America, Africa, and Asia.

Remarkable is the late entry into North Africa, which was closed by Berber tribes. The scientific branch of Napoleon's Egyptian expedition (1798-1802) gained valuable results, which were published in an exemplary travel book. The conquest of Algeria since 1830 is an important benchmark in the history of travel; after that time the exploration of the Sahara from Tripoli was accelerated. The hitherto existing history of discoveries had already recognized the great questions of exploration as the means to organize travel--the Nile, Niger, and Congo problems. Here it is remarkable that East Africa was well explored only in particular regions-- Abyssinia in 1837-1848 through the Abbadie brothers, Nubia in 1813-1814 through Burckhardt, and Kordofan in 1837-1838 through Russegger. In 1857 and 1858 Burton and Speke discovered lakes Victoria and Tanganyika. German geography took particular notice of Lichtenstein's trip in South Africa (1803-1805). Moritz von Beurmann, who participated in the search for the missing Vogel and in the process lost his life in 1863, had developed his own plan for the exploration of Africa. He recognized that travelers were constantly subject to the moment and in their reports could easily overemphasize accidental happenings. Therefore he proposed a network of scientific stations that could be cared for for four years by young European scientists, who should be relieved after having fulfilled their tasks. The accompanying European personnel would further develop these stations as centers for geographical research, feed themselves, and establish substations for further intensive research. Beurmann considered it his life's task to realize this aim.

In North America research was pushed forward by prospectors, who often had practical geographical experience. The United States early organized the exploration of its territory. In 1879 the Geological Survey was founded. There was also a Geographical Survey for the country west of the hundredth meridian. The Coast Survey had more tasks than is shown by its name. It commissioned J. G. Kohl with some literary tasks during his stay in the States. State expeditions opened up many areas. Friedrich Gerstäcker described the Ozark area for the first time, and Ritter's pupil A. Guyot inaugurated the research of the Appalachians and sent back hypsometric measurements. B. Möllhausen, who was supported by Humboldt, joined western expeditions as an artist. Beside the great participation of geologists, the importance of the journeys to members of the German nobility is remarkable.

After Humboldt's journey Middle America and parts of South America were neglected. On the other hand, von Eschwege did some excellent work in Brazil.[27] He also became the leader of the following German travelers: Martius, Spix, von Langsdorff, and Olfers. Maximilian II of Bavaria facilitated the expensive publication of the travel accounts of Spix and Martius, which were unparalleled at the time. As a further fruit of this journey the Flora Brasiliensis appeared, one of the great works of natural scientific literature. This report was only surpassed by Eduard Poeppig's art of representation. He was first a scientist but then proved his geographical gifts in consummate travel descriptions and was recognized and appreciated by many geographers. This research immediately became important to scientific geographers. Humboldt, like Goethe, received geological insights from von Eschwege. Ritter received accounts from Ignaz von Olfers, whom he, like Humboldt, had provided with instructions. Travelers such as J. G. Kohl and M. Wagner were traveling almost constantly and got to know the most different parts of the world. Always one can prove connections to Ritter and Humboldt, who analyzed the results of the new research with its innumerable details in the grand display of their geographical descriptions.

In Asia the research of the early 19th century was concentrated in the areas between the Far East and India. Here a type of political traveler developed. Englishmen, and later Russians, built a network of consulates, which also facilitated the tasks of the travelers of other nations. F. R. Chesney's large-scale steamship expedition to the Euphrates and Tigris (1835-1837) and P. A. von Chichachev's and M. Wagner's important journeys to Asia Minor and the Near East are examples of the enterprising spirit that advanced research. The connection of German and Russian travel research was so close that it led to great successes. This work toward a common aim was particularly strengthened through Humboldt's 1829 journey to the Urals and the Chinese border. Humboldt's enterprise was an inspiring example, but the personality of the German scientist also gave new inspiration to the Russian efforts. The methods of Alpine research from Humboldt to Wagner have trained the eyes of many travelers and resulted in the fertility of their observations in foreign countries.

Thomas G. Montgomerie (1830-1878), a leading officer of the Trigonometrical Survey in India, worked in land survey since 1852. For ten years he worked alone on the survey of Kashmir. The idea of sending Pundits [native Indian explorers] into the difficult to reach areas of Asia probably came from him. He instructed able natives in the use of compass and sextant but did not teach them how to evaluate their observations, probably in order to protect himself against possible falsifications. The Pundits had considerable success in the years between 1856 and 1880. The most famous of them, Naing Singh, helped the Schlagintweit brothers as early as 1856 in Kashmir and Ladakh, and in 1865 he traveled to Tibet and talked to the Dalai Lama in Lhasa. The English often kept the names of the Pundits and their results a secret for political reasons but also to protect their emissaries from acts of revenge. The achievement of these men was very important and should be more appreciated in the history of travel.

Remote Australia received its name in 1814 from Flinders. Only the coastline was known. The opening up was almost solely the effort of the English. Ludwig Leichhardt was prevented from the execution of his plans by his early death. He had studied in Göttingen and Berlin and in 1841 went to Sydney in order to prepare a larger enterprise by small excursions into the back country. In 1844-1846 he travelled from Brisbane into the unknown northeastern mountains up to the Gulf of Carpentaria. Thus he was one of the most successful travelers. In 1847 he tried to traverse the continent from east to west and disappeared in 1848.

The focus of all research was in Africa, but polar exploration equally captivated Mankind. Since the 16th century people believed in a great southern continent (Terra Australis). James Cook could not find it and concluded that the land could perhaps be situated around the South Pole. From this resulted A. Dalrymple's polemic against him, because Dalrymple unceasingly insisted on the existence of that land. F. F. Bellingshausen discovered and mapped in 1819-1821 some islands beyond the Antarctic Circle and the Alexander I coast, which since 1910 (Charcot) has been described as an island but certainly lies close to the Antarctic mainland.[28] For a long time one received only very sparse but important news from whalers and sealhunters, who pushed forward from the South Shetland and South Orkney Islands. After Humboldt and Gauss had inspired further journeys, the American Charles Wilkes, the Frenchman Dumont d'Urville, and the Englishmen Clarke Ross and F. R. Crozier pressed forward in the direction of the pole. Wilkes discovered in 1840 the coast that was named after him, and in 1841 Ross discovered Victoria Land with the volcanoes Erebus and Terror.

In the Arctic scientific interests pushed travel forward much earlier. One wanted to know the economic value of the marine fauna and to research the possibility of a North West Passage. Thus the Canadian Arctic Archipelago became known for the first time through Parry 1819-1822. Franklin followed in his steps with a great expedition of 145 members. After 1848 rescue expeditions were sent out as Franklin was missing. During the course of such an enterprise McClure and McClintock discovered the North West Passage in 1850-1854 and proved its economic and political worthlessness under the conditions of that time. E. K. Kane insisted in 1853 on the existence of an open northern polar sea.

3. First Stage of Modern German Geography

(1859-1869)

A strange fate had removed Humboldt and Ritter in one and the same year. So far the history of science has pointed out that these two great men could not really have absorbed the new ideas, as, for example, Darwin's On the Origin of Species was published only in their mutual death year. This is wrong and can be disproved by sources. Certainly, Humboldt and Ritter could not systematically take up and apply the new knowledge of genetics. Letters from Humboldt's estate have shown that he acknowledged the genetic studies in a modern way, promoted them, and also knew their consequences. Also Ritter was too much of a scholar to be able to overlook Lyellism, for example. In his works he often showed the most modern insights and used the word "genetic" in the modern sense, just like Humboldt. Therefore their deaths were not a catastrophe, and their work was not submerged and reborn anew after years. German geography after Ritter and Humboldt did not show signs of decay, nor was it purely derivative. Instead they dealt exclusively with the new--i.e., they tried to prove the fruitfulness of the idea of evolution for geography. Above all they did not just take after Humboldt, as perhaps some geographers themselves believed. In every geographical work of this time Humboldt's and Ritter's ideas are found in insoluble unity.

The results of the decade-long struggle for evolution are clearly demonstrated. Through the collaboration of geognosy and comparative anatomy, as mediated by paleontology, geology and evolution became the most modern natural sciences of the 19th century. It can be proven that the earth and its creatures have evolved. Geography was renewed in its physical part as "the morphology of the surface of the earth," as it is called in the title of Peschel's famous work, and Moritz Wagner gave new direction to physical and biological geography. In his Ausland Peschel published probably the first critical reviews of the Darwin work, whose contents were then criticized by Moritz Wagner in a geographical, i.e. spatial, sense. By finding numerous proofs for his migration law in nature, Wagner prepared the work of his only pupil, Friedrich Ratzel. The combination of geology and comparative anatomy can be seen with all geologists who opened up the wider development, in the form of ideas from

Peschel, but most obviously with Wagner, Ratzel, and Rütimeyer. Characteristically, their decisive works appeared in 1869, which introduced modern German geography and incorporated the idea of evolution into their concepts.

Wagner was an important traveler, and Rütimeyer, a thorough explorer of the Alpine world, should be placed with Scheuchzer and Saussure. Many enterprises during this time still derived from inspirations from Ritter and Humboldt. Wagner traveled through Panama and executed a plan by Humboldt by directing attention to the most commendable line to be drawn for the canal. Research in Asia Minor followed Ritter's ideas for a long time.

Geography was further developed in a few places. Wappaeus (since 1845 in Göttingen), C. Neumann (since 1863 in Breslau), and F. Simony (since 1851-1852 in Vienna) were successful. And the rare lectures of Barth, which he had given since 1863 as Ritter's successor in Berlin, attested to the rather important tendency toward morphology in this decade. He himself considered the study of relief his favorite subject and often met with the geologist Beyrich, Richthofen's teacher, who inspired his natural scientific thinking. Barth became the president of the Berlin Geographical Society and as a professor conducted annual excursions. Perhaps he could have become Ritter's heir, as he opted increasingly for more geographical work and wanted to conduct a systematic geography of Africa and a natural and culture-historical description of the Mediterranean Basin, but he died in 1865.[29]

It is in this period that field research developed faster. Again it is remarkable how in connection with the opening of the non-European continents, the opening up of Europe itself is restricted to certain specific areas, where a research that is rich in tradition and quality further develops (Switzerland, Austria, the Alps, Thuringia, and France).

In Africa Speke discovered in 1862-1863 that the Nile flowed out of Lake Victoria, in 1862 Rohlfs began his successful Sahara journeys, 1861-1863 Bastian crossed Southeast Asia for the first time, in 1863-1864 Pumpelly travelled for the first time through northern China and parts of Manchuria, and in 1860-1862 F. von Richthofen accompanied a Prussian delegation which was supposed to make commercial treaties in the Far East, and he got to know the Philippines, Japan, and Siam. During this journey he met Franz Junghuhn in Java, and the latter explained the geographical individuality of the island to him. In 1867-1869 he participated with F. Zirkel and Clarence King in a great geological expedition to the American West. In 1864 Severtsov's travels in the Tien Shan introduced a new chapter in the exploration of Inner Asia. In 1868 Fritsche crossed the eastern Gobi. The years 1868/69-1870 constitute in many respects a new beginning, as for Reiss and Stübel, Nachtigal and Przheval'skii. With regard to travel research and scientific geography this decade constitutes not an end but a transition.

4. August Petermann and the Flowering of

Exploration Geography

Hermann Wagner was of the opinion that the deaths of Humboldt and Ritter were not immediately followed by a further period of boom of scientific geography, but by the "Petermann era."[30] Certainly this gifted organizer and cartographer belongs to the personalities who connect the epochs of geography, but his influence does not so much stretch over the science as over the travel in his time.

August Petermann was born April 18, 1822 in Bleicherode in Eichsfeld. His father, August Rudolf Petermann, was an actuary. The youth attended school in Bleicherode and from 1836 on went to the Gymnasium in Nordhausen. According to his mother's wishes he was supposed to study theology, but Petermann became a student in the School of Geographical Art (opened in April 1839) of Heinrich Berghaus, who considered him a foster son and in a five-year apprenticeship trained him to be a cartographer.[31] In 1841 he met Humboldt for the first time, and the latter entrusted him with the drafting of the map for his work Asie centrale.[32] Humboldt and Ritter were also polestars for Petermann. From April 1844 he taught at the Berghaus school but left a year later, in October 1845, to become an assistant at Alexander Keith Johnston's cartographic institute in Edinburgh, where he collaborated on the English edition of Berghaus's Physical Atlas, which was by no means a mere imitation. He had already done a great deal as a collaborator on the original work, and his contribution to the English edition was considerable. A number of maps, for example the geological sheets of the first and fourth sections, he considered as his own work.[33] "Through his collaboration on two great physical atlases Petermann became a practical representative of Humboldt's ideas."[33] At that time his first publications appeared.

In June 1847 Petermann went to London and became a member of the Royal Geographical Society. In the early 1850s he was one of their secretaries but only for a brief time. Here his views enlarged mainly through personal acquaintances. On January 1, 1852 he opened his own geographical institute for the "sketching, etching, lithographing, and coloring of land and sea maps, geological and relief maps, plans and ground-plans."[33] Soon he asked for more space for his institute, a sign that it was developing favorably. On May 3, 1852 he enrolled his first pupil, E. G. Ravenstein from Frankfurt. Upon the intercession of the Prussian ambassador, von Bunsen, he was allowed to call himself "physical geographer and engraver on stone to the Queen." He soon realized that the prestige and influence of von Bunsen could help him. Through von Bunsen, who was a friend of Humboldt and Friedrich Wilhelm IV of Prussia, Petermann tried to realize his ideas. In 1849 he met the missionary James Richardson, who had already in 1845-1846 executed a much noticed Sahara journey and who had now received permission for a further expedition. Doubtless it is of great merit that Petermann succeeded in preparing the way for Barth, Overweg, and Vogel through collaboration with Bunsen, Humboldt, Ritter, and the English authorities. He had the confidence of the English, had incorporated English habits into his life-style, and was officially asked for advice. During this time in London the seed of his later success was already growing. Already during this time he could advertise in English journals that he could correct maps according to the results of travels.

Thus it was a good move when Perthes succeeded, after some initial resistance, in bringing August Petermann in August of 1854 to Gotha, where he gained his greatest successes. For years he was occupied with the cartographic aspects of Barth's travel works, which he finished in 1857-1858 as one of his most important works. In this time travel literature was multiplying, and he rightly judged the circumstances when he continued the Geographisches Jahrbuch of Heinrich Berghaus in the form of his Geographische Mitteilungen. The first number appeared on February 15, 1855. Shortly afterward, Petermann's Geographische Mitteilungen was recognized as the main journal of exploration geography; it now carried the results of research travels. From 1856 Dr. E. Behm was a co-editor and unnoticeably but steadily donated all his energy to the great aim of the publisher, to make a German journal the leading journal of geography. After 1860 the Ergänzungshefte of the Mitteilungen allowed for more detailed reports of the more important travel results. Petermann himself edited 23 volumes and 56 Ergänzungshefte and further exercised influence on the workings of geography through the choice of his collaborators. Moritz Wagner, for example, owes much to Petermann, and it is surely a sign of personal tact that Wagner never had to mention this.

After his move to Gotha Petermann chiefly pursued the exploration of Africa, without, however, losing sight of the other focal points of travel research. He immediately gathered pupils around him and brought new life to the Perthes institute.

Viewed historically, Petermann followed Rennell, whose work he got to know in England and continued by himself when he succeeded in insuring Barth's and Overweg's participation in an English expedition. Thus, apart from the economic and commercial-political motives, the scientific value of this journey was assured. For fifteen years he was busy with the organization and above all the cartographic evaluation of this expedition. In London he defended all attacks against the German travelers, continuously published partial results, and saw that the enterprise did not fail after Richardson's death. Already during the journey he showed his contemporaries how the map of Africa slowly became more complete, and his 16 maps for Barth's great travel work mirrored the results in detail. Often Barth's statements were corrected after comparison with other sources. Since these years Petermann's name became inseparably linked with the exploration of Africa. He mastered the details, knew best how to pose questions, and accordingly guided travelers. It was his idea to take along a collapsible boat to travel on Lake Chad on Barth's journey, which allowed Overweg to make a noteworthy accomplishment. His energy also made possible the "Pleiad" steamship expedition under Baikie in 1854, which discovered the merging of the Benue into the Niger. After Overweg's death Petermann evaluated his astronomic determinations, and he arranged to have Eduard Vogel sent to Barth as an astronomer. When Vogel was missing, Petermann organized the rescue expeditions from Neimann to Nachtigal, who finally brought back the ultimate explanation in 1873. For this purpose he organized a permanent commission with von Seebach as the director and Perthes as the treasurer, while he himself was content to take the office of secretary. Ten thousand talers were collected and a "German Rescue Expedition" was equipped, which was excellently prepared scientifically. He also facilitated the subsequent journeys of Moritz von Beurmann, G. Rohlfs, and Gustav Nachtigal, and he did not hesitate to go out and actually beg for money for travelers. Karl Mauch also confessed that without Petermann's aid his successes would not have been possible.

Petermann was the greatest expert on the history of discovery in Africa. He knew many travelers personally and probably knew all of them by correspondence.

Without his intervention we Germans could probably have helped only indirectly with the exploration of the polar regions. In his politically disunited country he succeeded in bringing Germans together in common expeditions. Political historians tend to overlook the fact that Petermann helped to exercise a unifying influence in a politically divided, particularistic time. His expeditions did not work under the flag of a single state but under the flag of all Germans. The accomplishment of the travelers he sponsored gained the enthusiasm of the German youth and awakened a united German consciousness. He used the commercial-political interests of other powers—if the direct way was not possible—to advance the scientific aims of the geography of his not yet unified country.

He prepared the ground for polar exploration in Germany by referring to the geographical, meteorological, geophysical, geological, oceanographic, and nautical motives. Already in London he was occupied with the fate of the Franklin expedition in a literary way, and from Gotha he tried to further inspire the English for polar travel. Finally he also transferred these ambitions to his homeland. He awakened enthusiasm, but even the most modest preconditions were lacking. Petermann had to charter a British ship because he could not get hold of a German one. But the English offered some difficulties, and in 1865 the enterprise failed because of the intrigues of British sailors, even though it had some precise aims: the exploration of Gillis Land, the coal resources of Spitzbergen, and the northward continuation of the Gulf Stream.

After some laborious preparations, Petermann then asked Captain Karl Koldewey to lead the first German expedition to the north polar regions. Koldewey had sailed the seas from 1853 to 1866, had been to North Cape and Archangelsk, and then from 1867 had studied mathematics, astronomy, and physics in Gottingen. According to Petermann's theories one should try to press forward in a northerly direction up the east coast of Greenland—i.e., to continue the research of Scoresby and Clavering. On May 24, 1868 the "Germania" left from Bergen, pushed forward along the eastern coast of Greenland to latitude 76°, and returned in September 1868. Koldewey sailed around the northern coast of Spitzbergen and for the first time surveyed the Hinlopen Strait. Finally he pushed beyond Spitzbergen to latitude 81°. In 1869 the second expedition began under Koldewey, who in the meantime had finished his studies, and Hegemann. Aboard the two ships, "Germania" (Koldewey, Dr. Copeland, J. Payer, and Dr. Boergen as astronomer) and "Hansa" (Hegemann) were several scientists with specialized backgrounds. The "Germania" wrecked below latitude 75°. The "Hansa" landed at Sabine Island, discovered and explored the Franz Joseph Fjord (Copeland and Payer). The ship reached 75°31', although the members of the expedition pushed forward on sleighs up to 77°1'. Islands and coastlines were mapped and new insights were gained into the nature of glaciers in Greenland. In 1871 Koldewey became an assistant in the naval observatory in Hamburg and, under the direction of Dove, worked out the meteorological and hydrographic results of the expeditions. Petermann also contributed to the scientific evaluation of these travels. When he was planning this enterprise he had to suffer from the small-minded attitudes of the separate German states. But he shook up the Germans and made his enterprise a concern of the people. The foundation of the first German North Sea Fishing Society was connected with it. Even after the foundation of the Empire, Petermann was granted no support; the great German polar research that he envisioned miscarried. Therefore he gave his help and experience to other countries: to the Frenchmen Lambert, O. Pavy (who took up Lambert's plan in 1871), Ambert, and Mack. He influenced the Swedish expeditions of 1868 and 1872 and saw A. E. Nordenskiöld go, according to his plan, in 1875 from Tromsö to the Siberian coast. Thus the search for the North East Passage had again become a vital problem.

Petermann got the British government to resume its polar research. After approving the way the English suggested, Nares explored with two steamers from 1874 to 1876 and at 82° reached the northernmost point of any ships thus far.

Petermann directed the researches of the Russians, who appreciated his advice very much, to the exploration of the Arctic Ocean (1869-1870) and called for the establishment of meteorological stations in the north and the exploration of the northwestern coast of Novaya Zemlya. This resulted in Aleksandrovich's expedition, of which Middendorf gave some valuable information to Petermann. Through the journey of Chekanovskii (1873) they tried to improve the former results. Also in Holland and Italy Petermann incessantly tried to encourage the equipping of new expeditions.

His thought of interesting the numerous Norwegian whalehunters in scientific tasks was a program by itself. Thus, for example, following his wish, the Norwegian government offered premiums to the ship captains for all meteorological observations. Following his advice, they went in increasing numbers toward Novaya Zemlya and the Kara See from 1868 to 1870. Their hints induced Nordenskiöld's journey of 1875. In our time people are hardly aware any longer of how one had to advance step by step in order to assure greater successes like Nordenskiöld's North East Passage. In Germany Rosenthal could be convinced to employ his sealhunters for scientific tasks at his own expense. Petermann's Mitteilungen collected the results. Justifiably Petermann could write in 1871 that Germany was the scientific focal point of polar research. It is little known that he coined many placenames on his maps--e.g., Bastian Island, Perthes Island, Berghaus Island, and Behm Island.

Again and again the word "incredible" returns when one considers Petermann's life. He published 226 scientific contributions and essays and 531 atlases and single sheets. His personal correspondence with governments and travelers is to be considered as part of his work.[34] He succeeded in planning and directing the African and polar research of his time. Furthermore he helped individual travelers such as M. Wagner, von Seebach, Leichhardt, Mauch, and E. Holub. Many researchers, for example Semenov, had personal contact with him. He was tirelessly active as "father of expeditions" and as a "beggar for research travelers." He was the teacher of the most important cartographers--E. G. Ravenstein, Bruno Hassenstein, Ernst Debes, Ludwig Friederichsen, and Arnim Welcker. In 1866 he had proposed a great German Geographical Society for the Organization of German Travels. But this highly interesting plan also failed, although more than 1000 members were willing to join. Unfortunately, Petermann did not find anybody who could have become the director and also did not find the desire to realize a common German cause by the fragmented German states. It was painful to him that German travelers had to sell themselves to foreign countries and fell into misery when their health had suffered. He referred to the social problems that were connected with each research journey. He cited examples to open the eyes of Germans; it was all in vain. After the foundation of the Empire Bastian succeeded in founding a German Society for the Exploration of Equatorial Africa, in which Petermann also participated. It was certainly a great organizational mistake that he was not much more considered in Berlin and that his great experience was not better used. Again, as so often in our history, one was satisfied with a part without striving for the whole. An "African Society" was founded, although Petermann's plan was much more promising.

About Petermann there has been much speculation. He was forgotten astonishingly quickly. Only when Ratzel inspired his pupil Hugo Weller to pay homage to Petermann's accomplishments did the organizational talent of this man become clear. At present, however, one is often influenced by the judgment of Banse, who dismissed him as pretentious and as having imagined himself to have executed the journeys that he had planned.[35] Banse mostly got his knowledge from a letter of Barth that was written in a moment of resentment. In reality, Petermann helped with all his individual strength and fulfilled the great task of his time by bringing travel finally under the control of geography. This was his greatest task and the historical meaning of his lifework. His suicide may have been caused by several reasons, among others certainly by a marital crisis and by overwork, perhaps also by the realization that the new capital, Berlin, was about to surpass the importance of Gotha to geography and thus his own position as an organizer of travel. Since 1875 he suffered from such moods. But all that can probably only partly explain why he committed suicide on September 25, 1878.

5. Geography under the Influence of Ratzel

and Richthofen (1870-1905)

Modern German geography developed as the morphology of the earth's surface, under the aspect of the theory of evolution. At first after 1869 there were lacking great personalities, who had been customary in the past, but they slowly developed. Peschel taught since 1871 in Leipzig, but he died in 1875. Perhaps his death saved him from being slowly forgotten, as Heinrich Berghaus had to face. There was by no means a gap, as for example Friedrich Simony and Alfred Kirchhoff as teachers ensured a great tradition. After the foundation of the Empire a central influence came to geography. Geographical professorships came into being, but there were no geographers, and so one had to call in scholars from other fields who, for the most part, deserve recognition for enlarging geography to a discipline. Kirchhoff was of great importance. He was a born teacher and soon became known along with Humboldt and Ritter. He combined rhetorical skill and pedagogical talent with methodological abilities in regional geography. He conducted exercises that nearly exclusively served the education of teachers, because he understood that the position of geography in the universities had to be backed by work in the schools. He maintained his position well and became one of the most successful teachers of geography in his time. He was by no means just a kind of school geographer within the university, but he was also a person rich in ideas and spirit, whose methodological inspirations have endured to the present. He supported Sven Hedin and thus made possible the first scientific work of the last great overland traveler. In a certain way, as Zeune once did, he prepared the way for two great geographers: Friedrich Ratzel and Ferdinand von Richthofen.

Ratzel was a pupil of M. Wagner, who descended from Humboldt and Ritter and revered both as masters. Richthofen had attended Ritter's lectures, but he was not inspired by him, as he, who belonged to the generation of J. Fröbel, wanted to work in a more scientific way. Out of shyness he did not approach Humboldt, but he later acknowledged Ritter, whom he often quoted with praise in his great work on China. Silently he always took Humboldt for an example. After a thorough geological education, both Ratzel and Richthofen became geographers through travel research, the former more through Moritz Wagner, the latter through Junghuhn and the researchers in the American cordilleras. French geographers justifiably called Ratzel's works in physiological geography important. Richthofen was the first to complete a modern geomorphology, but he was also a great regional geographer. Ratzel was the modern leader of anthropogeography, following the work of M. Wagner, and was above all a completely rounded geographer who, like Richthofen, was not at all tied to one particular trend within the science. One should not construct any insurmountable contrasts between them. In a certain sense we can compare them to Ritter and Humboldt, to whom both certainly owed something, although to different degrees; but that does not justify the view that, following Ritter and Humboldt, there had been two totally different branches of geographical tradition during the 19th century.

Richthofen's colloquium became the advanced school for future research travelers and geographers. Sven Hedin, Ernst Tiessen, Erich von Drygalski, Georg Wegener and others came out of his school. The borders between travel research and geography became blurred, as now science became forever the basis of travel. Ratzel's influence reached abroad; for example, J. Brunhes, E. de Martonne, and Miss Semple studied with him in Leipzig.

Adolf Bastian was perhaps the best expert on the world during this time. A pleasant personality, almost without a European sense of time! He was constantly traveling but nevertheless published each year stylistically impenetrable but brilliant works. Despite a strong ethnological tendency, Bastian had very close relations to the geography of his time. He graduated from Berlin in geography and then became director of the Berlin Geographical Society, which after the foundation of the Empire enlarged its influence, not least through its journal. He was the instigator of the German Society for the Exploration of Equatorial Africa and as its director traveled along the west coast of Africa in 1873 and founded the base station for travel research in Chinchonxo. Through grants and aid of the German Empire this society was able to equip travelers (Güssfeldt, Falkenstein, Pechuel-Loesche, Lenz, von Mechow, Lindner, Sagan, and Pogge).

Many an unfortunate later development was rooted in this imperialistic epoch. In 1884-1885 the German Empire became a colonial power. It brought about a special field that in a typical way was opened to German geography. The zoologist L. Schultze Jena became a geographer in the process. S. Passarge, W. Behrmann, and Fritz Jaeger (discoverer of the highland of giant craters in East Africa) are men who manifest for us the interconnection of field research and geography in these areas. The geographer, publisher, and research traveler Hans Meyer executed a world tour in the '80s; he traveled in South Africa in 1886 and in 1887, 1889, and 1898 explored Kilimanjaro, whose summit he climbed for the first time with Purtscheller. He became one of the most influential inspirers of German colonial geographical work.

Although Ratzel and Richthofen undoubtedly had a total view of geography, it was not characteristic of their time. There was a danger of geomorphology, as the most developed branch of the science, becoming independent. The brilliant work of the young Albrecht Penck, Morphologie der Erdoberfläche (1894), demonstrates this. Alfred Hettner, the first to devote himself entirely to the study of geography, averted this danger with his methodology. Strongly inspired by Kirchhoff, he drew attention to regional geography and brought the advanced morphology and anthropogeography back into the unity of the science.

The deaths of Ratzel (1904), Richthofen (1905), Reclus (1905), Bastian (1905), and Kirchhoff (1906) meant the end of this epoch. Whereas Ritter and Humboldt at the end of their days considered the further development of geography with sorrow, Ratzel and Richthofen, already in their lifetime, witnessed new developments that would last for nearly half of the new century, until the deaths of Hettner (1941) and Penck (1945).

Therefore, what we call the geography of the 19th century comprises approximately the years from 1799 to 1905. Centuries are for historians not physical unities of mechanical counting but reference concepts formed by historical experience.

The link between travel and scientific geography became more visible in this epoch. Already the year 1869 with its turn toward the morphology of the earth's surface was the result of an historical process and of the scientific evaluation of travel. The mere thought of the Führer für Forschungsreisende (1886) by Richthofen, of Penck's Morphologie (1894), and of the reports of numerous, now forgotten men may support this. Elisée Reclus's anarchist working technique did not make much use of quotations and thus the more brilliantly gave his vast material unity, as in the great regional geographical production of the Nouvelle géographie universelle in 19 volumes (1876-1894). The parallel German enterprise, Kirchhoff's Unser Wissen von der Erde (since 1885) unfortunately remained only a fragment. But Richthofen's China (1877-1885), Ratzel's Die Vereinigten Staaten von Nordamerika (1878-1880), and Penck's works in Kirchhoff's regional geography of Europe (1885-1889) on the German Empire, the Netherlands, Belgium, and Luxemburg represent the viewpoints of geography toward a discovered world that also had to be disclosed scientifically. Without the cordilleran research there would be no geomorphology. But the geographical explanation of this important area is an accomplishment of German science. Characteristic is the increasing participation of geologists in travel research. Travelers became increasingly dependent on scientific preparation and were brought to their task according to plans through geography. Most geographers of the time were inspired by Richthofen and Ratzel and may also have studied with the two.

During this time there were even in Europe regions that were not mapped. Kurt Hassert, for example, built his geographical research in Montenegro and Albania on personal cartographic surveys. Very important geographically was the continuation of the research started by H. Barth on the geographical unity of the Mediterranean area by Theobald Fischer and Alfred Philippson, because the areas that were treated previously allowed for especially thorough research and permitted definitive regional geographical masterpieces. In an epochal work Penck and Bruckner summarized the Alpine research of their predecessors and attained great progress, especially in glacial morphology. Hettner had trained his eyes through very productive travel in the Colombian Andes and wrote his dissertation (1887) on Saxon Switzerland. The literary art in the presentations of Theobald Fischer, Friedrich Ratzel, and Joseph Partsch is impressive.

In Africa, research became politicized, especially through the ambitions of the great colonial powers into whose hands all researchers soon played consciously or unconsciously. That certainly happened to the German travelers (Nachtigal, Rohlfs, Schweinfurth, and Emin Pasha), at least, but it lies deep in the peculiarities of the imperialistic era. Rohlfs, Nachtigal, and especially Schweinfurth received many inspirations from the exploration geography of the time. Rohlfs, the most successful, certainly personified more the type of the mere explorer, but he also tried to present scientific work, among other ways by taking along Karl A. Zittel and Paul Ascherson. In 1869 Nachtigal, through his visit to the Tibesti

Mountains, succeeded in the greatest discovery of the Sahara. His report is as important as his accomplishments as a traveler. Emin Pasha (Dr. Eduard Schnitzer) deserves great credit for the exploration of the Egyptian equatorial province but was unfortunately incredibly downgraded because he was Jewish.[36] Nevertheless he pursued aims that transcended imperialist colonialism, namely the fight against the slave trade and the civilizing of Africa. He was equally great as a discoverer and explorer and constantly tried to maintain contact with scientific geography. The Mahdi uprising and H. M. Stanley, who definitely executed a political mission, brought about Emin's fate. Stanley had rehearsed the "rescue" of Emin like a play, while the German rescue expedition of Carl Peters was impeded.[37] A closer examination of these events also led to a critical evaluation of Stanley's accomplishments. He was greatly indebted to the Arab leader Tippu Tib, whom he probably cheated in a conscienceless fashion. This is incidentally closely connected with the social avoidance of Stanley in England, as is presumably also the murder of Emin, which was commissioned by Tippu Tib, whom Stanley had blackmailed. This social avoidance is hardly appreciated by German research. Georg Schweinfurth deserves much recognition for the exploration of Northwest Africa and the Sudan; his scientific thoroughness surpassed that of most contemporary researchers. He discovered the Uele River, saw the first pygmies, and proved the existence of okapis. The outstanding personality in travel research in Southern Africa was David Livingstone, who solved the mystery of the great rivers, conducted scientifically considerable examinations, and as a researcher stands high above travelers such as Stanley, who wanted to be first mainly out of ambition in a time when scientific ability already had to legitimate and lend wings to the pure lust for adventure. Carl Mauch drew the first maps of the South African republics and discovered in 1868 the goldfields of Tati and in 1872 the Kaiser Wilhelm Goldfield as well as the ruins of Zimbabwe. Mauch deserves great credit, along with H. von Wissmann, Serpa Pinto, and Emil Holub.[38]

It was unfortunate that after Petermann's death there was little scope for such enterprising personalities as von Wissmann, C. Peters, and C. Mauch, who languished as a railway clerk.

In 1876 Leopold II of Belgium founded the "Association africaine internationale." They set as a task the exploration of inner Africa and intended to civilize the natives and fight the slave trade of the Arabs. Therefore Stanley founded stations on the Congo in 1879 and made treaties with the chiefs of the area. Out of this came in 1885 the Congo State, at first the private property of Leopold and after 1908 a Belgian colony. In 1884-1885 H. von Wissmann and others explored the Kassai region for the King.

This action of the Belgian king caused the formation of a new African Society in Germany on April 29, 1878, which after combining with the German division of the Belgian international "Association" replaced the older Society for the Exploration of Inner Africa. The state and most geographical organizations supported this society, which could work systematically and could send out several travelers: O. Schütt, M. Buchner, P. Pogge, and H. von Wissmann.

In Asia it was especially Russian travelers who promoted the exploration of Siberia and Inner Asia. Semenov organized many journeys and sketched the aims of their research. From 1885 D. N. Anuchin served as the first professor of geography in the University of Moscow since 1847. To him modern Russian geography is indebted for its foundation. Characteristically this important scholar dedicated his first lecture to the "evolutionary history of geography" because he recognized the necessity of historical scientific thinking. Anuchin thoroughly occupied himself with the methodology of Richthofen, Ritter, Peschel and others and trained a great number of geographers. The possibility of further travel in his own country and the collaboration of geology and other disciplines gave Russian geography an unmistakable character. V. V. Dokuchaev, the most important predecessor of K. D. Glinka, recognized the necessity of a geographic division into zones for science and practice. In the foreword to his Die Typen der Bodenbildung [1914, translated into English 1927] Glinka himself stated that the idea of the dependence of the soil on the climate, relief, vegetation, and bedrock is derived from the views of Dokuchaev. V. A. Obruchev was a geologist by training; he traveled with Potanin and also became well known in Germany. Since 1870 N. M. Przheval'skii crossed Inner Asia on four journeys, discovered the Tarim and the Lop Nor, explained the mountain structure of those regions, and was the first to cross the Humboldt and Ritter Mountains as well as the Nan Shan. He crossed the Gobi three times, found wild horses and camels there, and accomplished unparalleled feats, which then inspired Sven Hedin, who thoroughly analyzed the journeys of this great Russian. Hedin solved the Lop Nor problem, crossed the Taklamakan for the first time, traveled around the Tarim Basin, and discovered the source of the Brahmaputra and the Transhimalaya (1894-1908). Lhasa could be reached neither by Sven Hedin nor by Przheval'skii. After the Pundits had communicated the first reports, a Russian-trained Buryat succeeded in pushing forward to the mysterious city.

In the Arctic travels were introduced by the "Tegethoff" expedition of Payer and Weyprecht (1872-1874). Both were experienced travelers. Weyprecht, the scientifically most educated German polar researcher, wanted to set up permanent observation stations in the northern regions. The Tegethoff froze north of Novaya Zemlya. Nevertheless the researchers succeeded in discovering the Franz Joseph Land archipelago. A. E. von Nordenskiöld led all Swedish polar expeditions since 1868. Inspired by Petermann, he pushed forward to the Kara Sea (1875) and again proved that contrary to the beliefs of that time this sea is navigable in certain months. In 1876 he reached the mouth of the Yenisey and sailed up the river as far as 71°. Thus the possibility of a trade connection between Europe and Siberia was demonstrated, but also the North East Passage of the "Vega" (1878-1879) under Nordenskiöld.

Further polar research was conducted in an exemplary fashion, especially by Nansen. After an excellent physical and scientific preparation he crossed Greenland in 1888-1889 and described the nature of the ice in the interior. Unlike anyone before him, he considered the construction of important pieces of equipment and became the teacher of all subsequent polar travelers. In 1893-1896 he conducted the "Fram" expedition based on theoretical considerations. He rightly put the reaching of the North Pole behind other scientific aims, which had to be approached first. He proved the existence of a deep polar sea, demonstrated its oceanic character, and as a zoologist and oceanographer enriched our knowledge. Nansen's personality directed the subsequent researchers to their scientific tasks. But even his example could not prevent much energy from being senselessly wasted because of ambition in reaching the pole and without consideration of scientific aims. In 1892 and 1898 Peary explored northern Greenland, and he was perhaps the first to reach the pole in 1909, or possibly Cook reached it already in 1908. The most successful polar explorer was Roald Amundsen; he repeated the North West Passage (1903-1906) and conducted magnetic researches at the same time, because Amundsen, who often stayed with Neumayer in the Hamburg Naval Observatory, tried hard to justify his travels with scientific work.

The systematic exploration of Antarctica began later. After the "Challenger" expedition by Nares in 1875, reports were received from whalers, who especially since the '90s sailed in the Antarctic seas, which were rich in fish, and brought home important information (e.g., Larsen). Their activity in turn inspired scientific expeditions. The German oceanic expedition of the "Valdivia" under Cuhn and the "Gauss" expedition under Erich von Drygalski, which was inspired by Neumayer, were important scientific accomplishments. Drygalski, Richthofen's pupil, discovered Kaiser Wilhelm II Land. In the period 1901-1904 Robert F. Scott twice wintered in the Antarctic and reached latitude 82°17'. He found petrified plants of Tertiary age and described the ice cap. In 1908 Shackleton wanted to reach the South Pole by sled, but he had to turn back 180 km from his goal. The South Pole was reached for the first time in 1911 by Amundsen and five weeks later by Robert F. Scott.

Amundsen thought at the time that he had finished the work and that there was nothing else to be done. Undoubtedly an epoch was closed, although only in our time are the last secrets being taken from the Antarctic through the technically highly advanced travel research. Thus the history of discovery has reached its goal--yes, in a certain sense is already completed. Not so the history of travel, which is as vast as the changing questions of geography as a science.

6. Tasks for the History of Travel

The occupation with the history of travel has a value of its own, but it also makes available to contemporary geography many facts that are beneficial to consider. Each regional geographer has to concern himself with the history of discovery and development of the area that he is treating; it would be foolish to entirely ignore it. The history of geography has to raise the occupation with travel to a special branch of study. Thus one can win back many hidden treasures.

In the evaluation of the abundant material in travel literature we have to take two courses: generally one has to strive to investigate the organization of travels through exploration geography and to present the history of travel in its gross outline. Furthermore, the history of translations, of the observational instruments, of the entire equipment including the techniques of recording through sketches and photographs have to be considered without getting lost in detail and have to be understood within the context of the development of scientific geography. In particular the exploration of separate countries has to be evaluated. Here the various travel reports--on Mexico, for example--have to be considered in cross-section in order to facilitate explicit comparisons in a strictly specified time space of the geographical present. This procedure will prove very fruitful in the future. Until recently, even important geographers granted only a small historical dimension to scientific geography, but this has been enlarged. The history of travel has the same tendency when it compares the geographic endowment of landscapes and regions--according to the abundance of sources--within a set timespan in order to explain the present state.

The representation of a connection between travel and geography is going to give some advantages. Thus C. Ritter worked the manuscript of the traveler W. Schimper on western Arabia into his Erdkunde, Scherzer took over reports from M. Wagner, and Humboldt's teacher, C. W. Dohm, published for the first time Engelbert Kämpfer's Geschichte und Beschreibung Japans (1777-1779), a fact that says more than the reference to Robinson-Campe, who did not personally influence him anyway. Von Eschwege gave the Russians geological descriptions of Brazil that inspired analogies in the Urals. Countless men who are not mentioned anymore will again have something to say to us when we take their works from the library and evaluate them for the history of travel.

In his "Cultural Geography" [Recent Developments in the Social Sciences, 1927] Carl O. Sauer mentioned that Humboldt's Kosmos is of interest to us today only as an object of the history of science, although we come back again and again to his observations in Mexico and South America as the most trustworthy materials on those countries at the beginning of the 19th century. It is the same with each regional geographical work. The representations of Reclus, Penck, or Ratzel will always retain their value, but they would retain it even more if we take our work more seriously in the history of science. One has to show present-day geography its origins, its constitution, and its direction with relation to the past.

Beck's Footnotes (with emendations)

1. By travel the author means all enterprises that have enlarged our knowledge of the earth's surface or are still enlarging it at the present time and not just discovery or research expeditions. Tourists, travel writers, and poets have also conducted journeys with geographical effects and should thus be considered in the history of travel.

2. Hanno Beck, "Methoden und Aufgaben der Geschichte der Geographie," Erdkunde, vol. 8, no. 1 (February 1954), 51-57.

3. Hanno Beck, "Entdeckungsgeschichte und geographische Disziplinhistorie," Erdkunde, vol. 9, no. 3 (July 1955), 197-204.

4. Arthur Kühn, Die Neugestaltung der deutschen Geographie im 18. Jahrhundert. Ein Beitrag zur Geschichte der Geographie an der Georgia Augusta zu Göttingen. ("Quellen und Forschungen zur Geschichte der Geographie und Völkerkunde," vol. 5) Leipzig: K. F. Koehler Verlag, 1939.

5. John C. Beaglehole, The Journals of Captain James Cook on His Voyages of Discovery, Vol. 1, The Voyage of the Endeavour, 1768-1771 (Hakluyt Society, Extra Series, No. 34) (Cambridge: Cambridge University Press, 1955). This work is expected to run to four volumes; see the review by Hanno Beck in Erdkunde, vol. 10, no. 2 (May 1956), 175-176. [4 vols in 5 parts, Extra Series, Nos. 34-37, 1955-1974]

6. A. Plott, "Friedrich Konrad Hornemann (1772-1801)," Naturwissenschaftliche Rundschau, vol. 8, no. 6 (June 1955), 245-246.

7. Hans Plischke, "Johann Friedrich Blumenbachs Einfluss auf die Entdeckungsreisenden seiner Zeit," Abhandlungen der Gesellschaft der Wissenschaften zu Göttingen, Philologisch-historische Klasse, 3rd series, no. 20 (Göttingen, 1937).

8. Helmut Preuss, "Johann August Zeune in seiner Bedeutung für die Geographie," Dissertation, University of Halle, 1950.

9. For an informative account of the development of the planetarium, see Heinrich Berghaus, Wallfahrt durch's Leben vom Baseler Frieden bis zur Gegenwart, Von einem Sechsundsechziger (Leipzig: H. Costenoble, 1862), Vol. 6 [sic--only 4 vols. printed?], pp. 80ff. See also Hermann Wagner, "Gothas Bedeutung für die Pflege der Astronomie und Geographie," in Heinrich Anz (ed.), Gotha und sein Gymnasium (Gotha: F. A. Perthes, 1924), 146ff., and August Beck, Ernst der Zweite, Herzog zu Sachsen-Gotha und Altenburg als Pfleger und Beschutzer der Wissenschaft und Kunst (Gotha: J. Perthes, 1854).

10. The author has traced the roots of the history of geography as a science, but he believes that further study of the sources is necessary before he can reach a conclusion.

11. In lectures at the universities of Bonn and Cologne in January 1955 on "Ergebnisse und Probleme der geographischen Wissenschaftsgeschichte," the author characterized the epochs of geography since 1750 in detail. In this article a general view of the state of geography in each epoch will suffice.

12. H. Preuss, op. cit., 21.

13. Richard Bitterling, "Alexander von Humboldts Amerikareise in zeitgenössischer Darstellung," Petermanns Geographische Mitteilungen, vol. 98, no. 3 (September 1954), 161-171; Hanno Beck, "Alexander von Humboldt," Schweizerische Monatsschrift Du 1955, No. 10, pp. 22-33.

14. Hanno Beck, "Carl-Ritter-Forschungen," Erdkunde, vol. 10, No. 3 (August 1956), 227-233.

15. Johann Eduard Wappaeus (ed.), Carl Ritter's Briefwechsel mit Johann Friedrich Ludwig Hausmann (Leipzig: J. C. Hinrich, 1879), 129.

16. There are some rewarding tasks here for the history of travel. First, pictures should be collected and registered. Many representations have been lost. The author will show the history and significance of these pictures in a separate work. Therefore, in the rest of the article he will refrain from further mention of these facts of great importance to geography. One can often gain important insights from the pictures. The material that is available for research is unlimited. Such illustrations have occasionally been published in some lavish journals, and scientific geography should try to influence these publications in the future in order to gain greater results from them than have been obtained until now. See, for example, Hanno Beck, "Alexander von Humboldt," op. cit.

17. In addition, see Josef Röder and Hermann Trimborn (eds.), Maximilian Prinz zu Wied (Bonn: Ferdinand Dummlers Verlag, 1954). See review by Hanno Beck in Erdkunde, vol. 9, no. 2 (May 1955), 167-168.

18. Reinhard Frenzel, "Malthe Conrad Bruun (Malte-Brun), Frankreichs bedeutendster Geograph im ersten Viertel des 19. Jahrhunderts. Ein Beitrag zur Geschichte der geographischen Wissenschaft," Dissertation, University of Leipzig, 1908 (Crimmitschau: R. Raab, 1908).

19. A. von Humboldt wrote in 1820 to Malte-Brun: "I share your opinion of the difficulties that are presented; in Paris it is easier to found an Academy than an active committee and a fund to support travel." J.-B.-M.-A. Dezos de La Roquette, Oeuvres d'Alexandre de Humboldt. Correspondance inédite, scientifique et littéraire (Paris: L. Guérin, 1869), Vol. 1, p. 212.

20. Further statements in Hanno Beck, "Moritz Wagner in der Geschichte der Geographie," Dissertation, University of Marburg, 1951.

21. Reise der österreichischen Fregatte "Novara" . . ., Vol. 1 (Vienna: Kaiserlich-Königliche Hof- und Staatsdruckerei; in Commission bei K. Gerold's Sohn, 1861), Appendixes 1 and 2.

22. Robert Avé-Lallemant, "Ein Besuch bei Alexander von Humboldt im Jahre 1856," Deutsche Rundschau, vol. 56, no. 6 (March 1930), 233-236.

23. It should be mentioned that Blumenbach inspired the theme of the famous Göttingen prize work of Karl Ernest Adolf von Hoff, Geschichte der durch Uberlieferung nachgewiesenen natürlichen Veränderungen der Erdoberfläche, Ein Versuch, 5 vols. (Gotha: J. Perthes, 1822-1841), and through it founded German actualism. Lomonosov had already promoted the actualist principle in the 18th century, and von Hoff had already advocated it by 1810, long before Lyell.

24. Further statements in Hanno Beck, "Moritz Wagner," op. cit.

25. Dora Fischer, "Peter Petrowitsch Semjonow-Tian-Schanskij," Die Erde, vol. 5, no. 1 (March 1953), 67-70.

26. Quoted by Lev Semenovich Berg, Geschichte der russischen geographischen Entdeckungen, ed. by Rolf Ulbrich (Leipzig: Bibliographisches Institut, 1954), 194.

27. Hanno Beck, "Wilhelm Ludwig von Eschwege und die klassische deutsche Geographie," Erdkunde, vol. 9, no. 2 (May 1955), 89-92, and Hanno Beck, "Ergebnisse der W. L. von Eschwege-Forschung," Zeitschrift für Hessische Geschichte und Landeskunde, 1956, 164-173.

28. L. S. Berg, op. cit., 11.

29. Gustav von Schubert, Heinrich Barth, der Bahnbrecher der deutschen Afrikaforschung (Berlin: D. Reimer, 1897), 169.

30. Hermann Wagner, "Zum 100. Geburtstag August Petermann," Dr. A. Petermanns Mitteilungen . . ., vol. 68, no. 3 (April-May 1922), 77-78.

31. Hanno Beck, "Heinrich Berghaus und Alexander von Humboldt," Petermanns Geographische Mitteilungen, vol. 100, no. 1 (February 1956), 4-16.

32. According to Ewald Weller, "August Petermann. Ein Beitrag zur Geschichte der geographischen Entdeckungen und der Kartographie im 19. Jahrhundert," Quellen und Forschungen zur Erd- und Kulturkunde, vol. 4 (Leipzig, 1911), 18; see also Hugo Ewald Weller, "August Petermann als praktisch-organisatorisch tätiger Geograph . . .," Dissertation, University of Leipzig, 1904 (Partly published by J. Perthes of Gotha, 1904).

33. E. Weller (1911), op. cit., 18.

34. Dr. Werner Horn of Gotha has started to work through this correspondence. He intends to publish it, but due to the amount of material it can only be done in the form of registers.

35. Ewald Banse, _Lexikon der Geographie_, 2nd ed., vol. 2 (Leipzig: C. Merseburger, 1933), 300. The biographical articles in this dictionary were written by Banse himself and contain many personal attacks.

36. Erich zu Klampen, for example, dismisses him in his book, _Carl Peters_ (Berlin: H. Siep, 1938), as a "Jewish adventurer," whereas Hans Plischke in his _Entdeckungsgeschichte vom Altertum bis zur Neuzeit_ (Leipzig: Quelle und Meyer, 1933) gives him just recognition and says about him: "He had great organizational abilities and a deep drive to do research."

37. The author himself has considered the life and work of Emin Pasha and is soon going to publish the results.

38. Hans Offe, _Carl Mauch_. _Leben und Werk des deutschen Afrikaforschers_ (Tübingen: Druck von H. Laupp, 1937).

APPENDIX

BIOBIBLIOGRAPHICAL NOTES TO ACCOMPANY THE PRECEDING ESSAYS

In this section I am supplying brief biographical and bibliographical notes on the authors of the essays and on the modern geographers (Humboldt, Ritter, et seq., but not the explorers) mentioned therein. My aim is to be selective, not exhaustive, but these data are sufficient to put the reader on the trail of more information. In the case of living geographers, I have given the titles of a few of their significant publications; for the deceased, such information can be found in obituaries.

Where pertinent, abbreviated references to the following general sources are given:

Allegemeine Deutsche Biographie. Leipzig: Duncker & Humblot, 56 vols., 1875-1912. (Cited as ADB)

G. R. Crone. Modern Geographers: An Outline of Progress in Geography since AD 1800. Revised edition. London: Royal Geographical Society, 1970. (Cited as Crone)

Robert E. Dickinson. The Makers of Modern Geography. New York and Washington: Frederick A. Praeger, Publishers, 1969. (Cited as Dickinson, Makers)

Robert E. Dickinson. Regional Concept: The Anglo-American Leaders. London, etc.: Routledge & Kegan Paul, 1976. (Cited as Dickinson, Reg. Con.)

Dictionnaire de biographie française. Paris: Librairie Letouzey et Ané, 13+ vols. since 1933 (Fascicule 80, Foncemagne-Forot, pub. 1976). (Cited as DBF)

Dictionary of Scientific Biography, ed. by Charles C. Gillispie. New York: Charles Scribner's Sons, 16 vols., 1970-1980. (Cited as DSB)

Eric Fischer, Robert D. Campbell, and Eldon S. Miller. A Question of Place: The Development of Geographic Thought. Arlington, Virginia: R. W. Beatty, Ltd., 2nd ed., 1969. (Cited as Fischer)

Thomas Walter Freeman. A Hundred Years of Geography. Chicago: Aldine Publishing Company, 1961. (Cited as Freeman)

Geographers: Biobibliographical Studies, ed. by T. W. Freeman and Philippe Pinchemel. London: Mansell, 6 vols. since 1977. (Cited as Geographers)

"Les Géographes français." Secrétariat d'état aux universités, Comité des travaux historiques et scientifiques, <u>Bulletin</u> <u>de</u> <u>la</u> <u>section</u> <u>de</u> <u>géographie</u>, vol. 81. Paris: Bibliothèque Nationale, 1975. (Cited as "Géographes")

<u>Great</u> <u>Soviet</u> <u>Encyclopedia</u>. Translation of third edition of <u>Bol'shaia</u> <u>Sovetskaia</u> <u>Entsiklopediia</u>. New York and London: Macmillan, 31 vols. since 1973. (Cited as <u>GSE</u>) (Note: I have altered the spellings of Russian names in the essays and notes to conform, wherever possible, to the <u>GSE</u>)

Preston E. James. <u>All</u> <u>Possible</u> <u>Worlds</u>: <u>A</u> <u>History</u> <u>of</u> <u>Geographical</u> <u>Ideas</u>. Indianapolis and New York: The Odyssey Press (The Bobbs-Merrill Company, Inc.), 1972. (Cited as James) (Second edition, 1981, co-authored by G. J. Martin. References here are to first edition.)

André Meynier. <u>Histoire</u> <u>de</u> <u>la</u> <u>pensée</u> <u>géographique</u> <u>en</u> <u>France</u> <u>(1872-1969)</u>. Paris: Presses Universitaires de France, 1969. (Cited as Meynier)

<u>Neue</u> <u>Deutsche</u> <u>Biographie</u>. Berlin: Duncker & Humblot, 13 vols. since 1953 (Vol. 13, Kre-Lav, pub. 1982). (Cited as <u>NDB</u>)

Dimitrii Nikolaevich Anuchin (1843-1923)

Russian geographer and anthropologist. Professor in Moscow University from 1884. Director of the subdepartment of geography in Moscow University, 1885-1923.

V. A. Esakov in <u>DSB</u>, vol. 1 (1970), 173-175; <u>GSE</u>, vol. 2 (1973), 186-187; James, 289, 295n; D. J. M. Hooson, "The Development of Geography in Pre-Soviet Russia," <u>Annals</u> <u>of</u> <u>the</u> <u>Association</u> <u>of</u> <u>American</u> <u>Geographers</u>, vol. 58 (1968), 263-265.

Edouard Ardaillon (1867-1926)

French geographer. Professor of geography, University of Lille, fl. 1893-1901. Rector of the Academy of Algiers from 1908.

E.-G. Ledos in <u>DBF</u>, vol. 3 (1939), col. 436.

Konstantin Ivanovich Arsen'ev (1789-1865)

Russian geographer, historian, and statistician. Professor in St. Petersburg University, c.1819-1821. Academician of the St. Petersburg Academy of Sciences, from 1836.

<u>GSE</u>, vol. 2 (1973), 369; James, 283; D. J. M. Hooson, "The Development of Geography in Pre-Soviet Russia," <u>Annals</u> <u>of</u> <u>the</u> <u>Association</u> <u>of</u> <u>American</u> <u>Geographers</u>, vol. 58 (1968), 255-256.

Ewald Banse (1883-1953)

Professor of geography in the Technische Hochschule of Hannover from 1933.

Fischer, 167-174; Obit. in Rivista Geografica Italiana, vol. 61 (1954), 80-81.

Heinrich Barth (1821-1865)

German explorer in Africa. Professor of geography in the University of Berlin, 1863-1865.

Löwenberg in ADB, vol. 2 (1875), 96-99; Klaus Schroeder in NDB, vol. 1(1953), 602-603; A. H. M. Kirk-Greene, ed., Barth's Travels in Nigeria (London: Oxford University Press, 1962), "Introduction," pp. 1-75; Heinrich Schiffers, ed., Heinrich Barth, Ein Forscher in Afrika: Leben-Werk-Leistung (Wiesbaden: Franz Steiner, 1967); Hanno Beck, Grosse Reisenden (Munich: Callwey, 1971), chapter on Barth, pp. 239-270, 405-408.

Adolf Bastian (1826-1905)

Curator of ethnography in Berlin Museum from 1868.

Dickinson, Makers, 63; Hans Plischke in NDB, vol. 1 (1953), 626-627; Robert H. Lowie, The History of Ethnological Theory (New York, etc.: Holt, Rinehart and Winston, 1937), chapter 4, "Adolf Bastian," pp. 30-38; obits. by Karl von den Steinen and Ferdinand von Richthofen in Zeitschrift für Ethnologie, vol. 37 (1905).

Henri Baulig (1877-1962)

French geographer. Professor of geography, University of Strasbourg, 1919-1947.

Dickinson, Makers, 238-239; Fischer, 231-241; Etienne Juillard in "Géographes," 119-131; Meynier, 59; E. Juillard and C. Klein in Geographers, vol. 4 (1980), 7-17; obits. by Juillard in Annales de géographie, vol. 71 (1962), 561-566, and Pierre Marthelot in Bulletin de la Faculté des lettres de Strasbourg (November 1962).

Hanno Beck (b. 1923)

German geographer. Professor of the History of Science in Bonn University. Author of numerous works in the history of geography, including Alexander von Humboldt (2 vols., 1959-1961), Grosse Reisende (1971), Geographie (1973), Carl Ritter, Genius der Geographie (1979), and Grosse Geographen (1982).

Augustin Bernard (1865-1947)

French geographer. Professor of geography and of North African civilization in the University of Paris, 1902-1933.

M. Prevost in DBF, vol. 6 (1954), cols. 48-49; Keith Sutton in Geographers, vol. 3 (1979), 19-27; Meynier, 31, 182; obit. by Marcel Larnaude in Annales de géographie, vol. 57 (1948), 56-59.

Raoul Blanchard (1877-1965)

French geographer. Professor of geography in the University of Grenoble, 1906-1948.

Dickinson, Makers, 234-236; Juliette Taton in DSB, vol. 2 (1970), 190-191; Paul Guichonnet and Jean Masseport in "Géographes," 133-144; Meynier, 38; Blanchard's autobiographical works, Ma jeunesse sous l'aile de Peguy and Je découvre l'universite (Paris: A. Fayard, 1961-1963); obit. by Jean Dresch and Pierre George in Annales de géographie, vol. 75 (1966), 1-5 (followed by bibliography of Blanchard's writings, compiled by Françoise Grivot, pp. 5-25).

Eduard Brückner (1862-1927)

Austrian geographer. Professor of geography in the University of Vienna from 1906.

Dickinson, Makers, 109; Osterreichisches Biographisches Lexikon, 1815-1950, vol. 1 (Graz: Verlag Hermann Böhlaus Nachf., 1957), 120.

Jean Brunhes (1869-1930)

French geographer. Professor of human geography in the Collège de France, 1912-1930.

Dickinson, Makers, 212-214; Y. Chatelain in DBF, vol. 7 (1956), cols. 554-555; Juliette Taton in DSB, vol. 2 (1970), 538-539; Fischer, 223-231; Freeman, 305; Mariel Jean-Brunhes Delamarre (Brunhes' daughter) in "Géographes," 49-80; James, 250n; Meynier, 67-70; H. Froidevaux in Larousse Mensuel, vol. 7, no. 247 (September 1927), 508, and vol. 8, no. 287 (January 1931), 594-595; Rudolf H. A. Cools, De Geografische Gedachte bij Jean Brunhes (Utrecht: Kemink en Zoon N. V., 1942); obits. by Emm. de Martonne in Annales de geographie, vol. 39 (1930), 549-553, and Camille Vallaux in La Géographie, vol. 34 (1930), 237-239.

George Chisholm (1850-1930)

British geographer. Lecturer in geography in the University of Edinburgh, 1908-1921, and Reader, 1921-1923.

Dickinson, Reg. Con., 93-95 (reprint of Ogilvie, below); Freeman, 306; Kenneth Maclean, "George G. Chisholm: His Influence on University and School Geography," Scottish Geographical Magazine, vol. 91 (1975), 70-78; M. J. Wise, "A University Teacher of Geography," Institute of British Geographers, Transactions, no. 66 (1975), 1-16; obit. by A. G. Ogilvie in Scottish Geographical Magazine, vol. 46 (1930), 101-104.

André Cholley (1886-1968)

French geographer. Professor of geography in the University of Paris, 1927-1956.

Dickinson, <u>Makers</u>, 249-252; J. Gras in "Géographes," 153-171; obit. by Pierre Birot in <u>Annales de géographie</u>, vol. 78 (1969), 129-130.

Paul Claval (b. 1932)

French geographer. Professor of geography in the University of Paris. Prolific author of works in human geography, including <u>Essai sur l'évolution de la géographie humaine</u> (1964, 2nd ed. 1976), <u>La Pensée géographique: Introduction à son histoire</u> (1972), and <u>Les Mythes fondateurs des sciences sociales</u> (1980).

Jovan Cvijić (1865-1927)

Yugoslav geographer. Professor of geography in the University of Belgrade from 1893.

Fischer, 298-304; Freeman, 307; Milorad Vasovic in <u>Geographers</u>, vol. 4 (1980), 25-32; James, 335-336; T. W. Freeman, <u>The Geographer's Craft</u> (Manchester: Manchester University Press, 1967), chapter 4, "Jovan Cvijić, A Reluctant Political Geographer," 72-100; obit. by Lucien Gallois in <u>Annales de géographie</u>, vol. 36 (1927), 181-183.

François de Dainville (1909-1971)

French historian of cartography. Ecole des Chartes, Paris from 1959; Ecole Pratique des Hautes Etudes from 1963 (Director of Studies in Western Historical Cartography from 1966).

Françoise Grivot in "Géographes," 197-198; obit. by M. de la Roncière in <u>Imago Mundi</u>, vol. 26 (1972), 71-74.

Wiliam Morris Davis (1950-1934)

American geographer. Professor of physical geography in Harvard University, 1890-1899, Sturgis-Hooper Professor of Geology, 1899-1912.

Crone, 48; Dickinson, <u>Reg. Con.</u>, 193-208; Sheldon Judson in <u>DSB</u>, vol. 3 (1971), 592-596; Fischer, 379-392; Freeman, 307; Robert P. Beckinsale and Richard J. Chorley in <u>Geographers</u>, vol. 5 (1981), 27-33; James, 351-365; Reginald Daly, "Biographical Memoir of William Morris Davis," National Academy of Science, <u>Biographical Memoirs</u>, vol. 23 (1945), 263-303; "Symposium on Geomorphology in Honor of the 100th Anniversary of the Birth of William Morris Davis," <u>Annals of the Association of American Geographers</u>, vol. 40 (1950), 171-236; Richard J. Chorley, Robert P. Beckinsale, and Antony J. Dunn, <u>The History of the Study of Landforms</u>, vol. 2, <u>The Life and Work of William Morris Davis</u> (London: Methuen & Co., Ltd., 1973); obit. by Kirk Bryan in <u>Annals of the Association of American Geographers</u>, vol. 25 (1935), 25-31.

Albert Demangeon (1872-1940)

French geographer. Professor of geography in the University of Paris, 1911-1940.

Dickinson, Makers, 231-234; Fischer, 213-223; Freeman, 308; Aimé Perpillou (Demangeon's son-in-law) in "Géographes," 81-106; Meynier, 69-109 passim; E. Franceschini in DBF, vol. 10 (1965), col. 963; obits. by Emm. de Martonne in Annales de géographie, vol. 49 (1940), 161-169, and Lucien Febvre in Annales d'histoire sociale, vol. 3 (1941), 81-89.

Roger Dion (1896-1981)

French geographer. Professor of geography in the Collège de France, 1948-1968. Author of numerous works, including Le Val de Loire (1934), Histoire de la vigne et du vin en France des origines au XIXe siècle (1959), and Histoire des levees de la Loire (1961).

Dickinson, Makers, 243-244; Meynier, 82, 107, 109; J. L. M. Gulley, "The Practice of Historical Geography: A Study of the Writings of Professor Roger Dion," Tijdschrift voor Economische en Sociale Geografie, vol. 52 (1961), 169-183; Raymond Chevallier, ed., Mélanges offerts à Roger Dion: Littérature greco-romaine et geographie historique (Paris: A. & J. Picard, 1974); obituary by Pierre Gourou in Journal of Historical Geography, vol. 8 (1982), 182-184.

Vasily Vasilievich Dokuchaev (1846-1903)

Russian geologist and soil scientist. Affiliated with St. Petersburg University, 1872-1897, professor from 1883. As director of the Novo-Aleksandriia Institute of Agriculture and Forestry, he established Russia's first subdepartment of soil science in 1895.

V. A. Esakov in DSB, vol. 4 (1971), 143-146; Esakov in Geographers, vol. 4 (1980), 33-42; S. S. Sobolev in GSE, vol. 8 (1975), 343; James, 287-289; D. J. M. Hooson, "The Development of Geography in Pre-Soviet Russia," Annals of the Association of American Geographers, vol. 58 (1968), 262-263.

Theobald Fischer (1846-1910)

German geographer. Professor of geography in the University of Kiel, 1879-1883, University of Marburg from 1883.

Dickinson, Makers, 96; Günter Glauert in NDB, vol. 5 (1961), 205-206; Obituaries and appreciations by Hermann Wagner in Petermanns Geographische Mitteilungen, vol. 56 (1910), 188-189, K. Oestreich in Geographische Zeitschrift, vol. 18 (1912), 241-254, Roberto Almagià in Rivista Geografica Italiana, vol. 18 (1911), 332-351, and A. Rühl in Geographische Zeitschrift, vol. 27 (1921), 29-33.

Lucien Gallois (1857-1941)

French geographer. Professor of geography in the University of Paris, 1893-1927.

Dickinson, Makers, 226-227; Fischer, 198-202; Freeman, 310-311; A. Meynier in "Géographes," 25-33; James, 247n-248n; Meynier, 30, 103; obit. by Emm. de Martonne in Annales de géographie, vol. 50 (1941), 161-167.

Francis Galton (1822-1911)

British traveler, writer, anthropologist, geographer, meteorologist, and eugenicist.

Norman Gridgeman in DSB, vol. 5 (1972), 265-267; James, 256-257; Karl Pearson, The Life, Letters and Labours of Francis Galton, 4 vols. (London, 1914-1930); T. W. Freeman, The Geographer's Craft (Manchester: Manchester University Press, 1967), chapter 2, "Francis Galton, A Victorian Geographer," 22-43; William Warntz and Peter Wolff, Breakthroughs in Geography (New York: New American Library, 1971), 151-180.

Emile-Félix Gautier (1864-1940)

French geographer. Professor of geography in the University of Algiers from 1902.

Marcel Larnaude in "Géographes," 107-118.

Pierre George (b. 1909)

French geographer. Professor of geography in the University of Paris from 1948. Prolific textbook writer, with books on various branches of human geography (population, urban, economic, energy, rural, etc.).

Dickinson, Makers, 244-246; Meynier, 153-154, 186-188.

Grove Karl Gilbert (1843-1918)

American geologist and geomorphologist. United States Geological Survey from 1879 (Chief Geologist, 1889-1892).

Ronald DeFord in DSB, vol. 5 (1972), 395-396; Freeman, 311; Preston James in Geographers, vol. 1 (1977), 25-33; W. M. Davis, "Biographical Memoir. Grove Karl Gilbert, 1843-1918," National Academy of Science, Biographical Memoir, vol. 21 (1926); Henri Baulig, "La Leçon de Grove Karl Gilbert," Annales de géographie, vol. 67 (1958), 289-307; Stephen J. Pyne, Grove Karl Gilbert, A Great Engine of Research (Austin: University of Texas Press, 1980) (revised version of Pyne's PhD thesis, University of Texas, 1976); obit. by W. C. Mendenhall in Geological Society of America, Bulletin, vol. 31 (1920), 26-45 (followed by bibliography of Gilbert's writings by B. D. Wood and G. B. Cottle, 45-64).

Konstantin Dmitrievich Glinka (1867-1927)

Russian soil scientist. Affiliated with the Novo-Aleksandriia Agricultural Institute, 1895?-1913?; Director, Voronezh Agricultural Institute, 1913-1922, Leningrad Agricultural Institute, 1922-1927.

Iu. A. Liverovskii in GSE, vol. 6 (1975), 429; James, 289.

Arnold Henri Guyot (1807-1884)

Swiss-American geographer. Professor of geology and geography, Princeton University, 1854-1884.

Albert Carozzi in DSB, vol. 5 (1972), 599-600; Fischer, 363-370; Freeman, 311-312; Edith Ferrell in Geographers, vol. 5 (1981), 63-71; James, 192-194; Leonard C. Jones, "Arnold Guyot et Princeton," University of Neuchâtel, Faculty of Letters, Recueil de travaux, no. 14 (1929); Leonard C. Johes, "Arnold Henry Guyot," Union College Bulletin, vol. 23, no. 2 (January 1930), 31-65; Robert Anstey, "Arnold Guyot, Teacher of Geography," Journal of Geography, vol. 57 (1958), 441-449; William Warntz and Peter Wolff, Breakthroughs in Geography (New York, 1971), 132-150; obits. by William Libbey in Journal of the American Geographical Society, vol. 16 (1884), 194-221, and Charles Faure in Le Globe, vol. 23 (1884), Mémoires, 3-72.

Wolfgang Hartke (b. 1908)

German geographer. Professor of geography in the Technische Hochschule of Munich from 1952. Author of numerous articles, with particular emphasis on social geography. See the bibliography in the festschrift given to him on his 60th birthday: Zum Standort der Sozialgeographie (1968).

Richard Hartshorne (b. 1899)

American geographer. Taught in the University of Minnesota, 1924-1940, and the University of Wisconsin, 1940-1970 (Professor since 1941). Author of studies in economic and political geography but best known for his methodological works: The Nature of Geography (1939) and Perspective on the Nature of Geography (1959).

Fischer, 435-441; James, 417-421.

Ernst Emil Kurt Hassert (1868-1947)

German geographer. Professor of geography in the University of Tübingen, 1899-1902, Handelhochschule in Cologne, 1902-1917, and Technische Hochschule in Dresden, 1917-1935.

Hans Lippold in NDB, vol. 8 (1969), 48-49; M. Reuther in Petermanns Geographische Mitteilungen, vol. 94 (1950), 89-92.

Andrew John Herbertson (1865-1915)

British geographer. Taught at Oxford University, 1899-1915 (Professor since 1910).

Crone, 42, 44; Dickinson, Reg. Con., 43-54; Fischer, 261-267; Freeman, 312; L. J. Jay in Geographers, vol. 3 (1979), 85-92; James, 263-265; "A. J. Herbertson Centenary Special Issue," Geography, vol. 50 (1965) (articles by E. W. Gilbert, J. F. Unstead, H. J. Fleure, and L. J. Jay); obit. by H. J. Mackinder et al. in The Geographical Teacher, vol. 8 (1915), 143-146.

Alfred Hettner (1859-1941)

German geographer. Professor of geography in the University of Heidelberg, 1899-1928.

Crone, 35-36; Dickinson, Makers, 112-125; Fischer, 106-113; Freeman, 312-313; James, 226-229; Ernst Plewe in NDB, vol. 9 (1971), 31-32; Richard Hartshorne, "Alfred Hettner," International Encyclopedia of the Social Sciences, vol. 6 (1968), 354-356; Maximilien Sorre, "A. Hettner et la géographie de l'homme," Annales de géographie, vol. 58 (1949), 338-339; "Alfred Hettner: Gedankschrift zum 100 Geburtstag," Heidelberger Geographische Arbeiten, no. 6 (1960); obit. by Heinrich Schmitthenner in Geographische Zeitschrift, vol. 47 (1941), 441-468.

Eugene Woldemar Hilgard (1833-1916)

American geologist and soil scientist. Professor of agricultural botany, University of California, 1875-1906.

George P. Merrill in Dictionary of American Biography, vol. 5 (1960), 22-23; Fred Slate in National Academy of Science, Biographical Memoir, vol. 9 (1920); E. S. Smith in Bulletin of the Geological Society of America, vol. 28 (1917); Hans Jenny, E. W. Hilgard and the Birth of Modern Soil Science (Pisa: Collana della Rivista "Agrochimica," 1961).

Alexander von Humboldt (1769-1859)

German geographer, traveler, writer, and naturalist.

Crone, 14-18; Dickinson, Makers, 22-33; Kurt-R. Biermann in DSB, vol. 6 (1972), 549-555; Fischer, 61-65; Freeman, 313; James, 148-164; Hanno Beck in International Encyclopedia of the Social Sciences, vol. 6 (1968), 545-546; Hanno Beck, Alexander von Humboldt, 2 vols. (Wiesbaden: Steiner, 1959-1961); Hanno Beck, "Alexander v. Humboldt-- der massgebende Forschungsreisende," Grosse Reisenden (Munich, 1971), 131-145, 397-398; Helmut de Terra, Humboldt (New York: Knopf, 1955); L. Kellner, Alexander von Humboldt (London: Oxford University Press, 1963); Karl Sinnhuber, "Alexander von Humboldt, 1769-1859," Scottish Geographical Magazine, vol. 75 (1959), 89-101; Oskar Schmieder,

"Alexander von Humboldt . . .," Geographische Zeitschrift, vol. 52 (1964), 81-95.

Douglas Wilson Johnson (1878-1944)

American geographer. Taught geomorphology at Columbia University from 1912 (Professor of geology, 1919-1944).

Dickinson, Reg. Con., 284-286 (reprint of Wright's obituary, below); R. J. Chorley in DSB, vol. 7 (1973), 143-145; Freeman, 314; J. K. Wright, "Wild Geographers I Have Known," Professional Geographer, vol. 15 (1963), 3; obits. by J. K. Wright in Geographical Review, vol. 34 (1944), 317-318, and A. K. Lobeck in Annals of the Association of American Geographers, vol. 34 (1944), 216-222.

Heinrich Kiepert (1818-1899)

German geographer and cartographer. Professor of geography in the University of Berlin from 1874.

Dickinson, Makers, 54; obit. by J. Partsch in Geographische Zeitschrift, vol. 7 (1901).

Alfred Kirchhoff (1838-1907)

German geographer. Professor of geography in the University of Halle, 1873-1904.

Dickinson, Makers, 96-97; E. Meynen in Geographers, vol. 4 (1980), 69-76; H. Steffen, "Alfred Kirchhoff," Geographische Zeitschrift, vol. 25 (1919), 289-302.

Johann Georg Kohl (1808-1878)

German geographer and historian of cartography and exploration. Librarian in Bremen from 1863.

W. Wolkenhauer in ADB, vol. 16 (1882), 425-428; E. Cammaerts, "J. G. Kohl et la geographie des communications," Bulletin de la Societe royale belge de géographie, vol. 28 (1904), 36-61, 110-132, 225-243; Thomas Peucker, "Johann Georg Kohl, A Theoretical Geographer of the 19th Century," The Professional Geographer, vol. 20 (1968), 247-250; Gottfried Pfeifer, "Man Sollte J. G. Kohl Nicht Vergessen!" Mensch und Erde: Festschrift für Wilhelm Müller-Wille . . ., ed. by K.-F. Schreiber and P. Weber (Westfälische Geographische Studien, 33) (Münster, 1976).

Halford John Mackinder (1861-1947)

British geographer. Reader in geography in Oxford University, 1887-1905, and Director of the School of Geography, 1899-1905; Professor in the London School of Economics, 1895-1925, and Director of the School, 1903-1908.

Crone, 37-42; Dickinson, Reg. Con., 35-42; Fischer, 258-261; Freeman, 316; James, 258-261; Edmund Gilbert in International Encyclopedia of the Social Sciences, vol. 9 (1968), 515-516; Edmund Gilbert, British Pioneers of Geography (Newton Abbot, England: David and Charles, 1972), 139-179; Brian Blouet, "Sir Halford Mackinder, 1861-1947: Some New Perspectives," University of Oxford, School of Geography, Research Paper, no. 13 (1975); Brian Blouet, "Sir Halford Mackinder as British High Commissioner to South Russia, 1919-1920," Geographical Journal, vol. 142 (1976), 228-236.

George Perkins Marsh (1801-1882)

American diplomat, writer, geographer, conservationist, and philologist. U. S. Minister to Turkey, 1849-1853; Minister to Italy, 1861-1882.

James, 194-196; David Lowenthal, George Perkins Marsh: Versatile Vermonter (New York: Columbia University Press, 1958); David Lowenthal, "George Perkins Marsh on the Nature and Purpose of Geography," Geographical Journal, vol. 126 (1960), 413-417; David Lowenthal in International Encyclopedia of the Social Sciences, vol. 10 (1968), 23-25.

Emmanuel de Martonne (1873-1955)

French geographer. Professor of geography in the University of Paris, 1909-1944.

Dickinson, Makers, 229-231; Robert Beckinsale in DSB, vol. 9 (1974), 149-151; Fischer, 180-186; Freeman, 308; Jean Dresch in "Géographes," 35-48; James, 251n; Meynier, 33, 41, 43, 45-48, 64-66; obits. by André Cholley in Annales de géographie, vol. 65 (1956), 1-14, and Max. Sorre in IGU Newsletter, vol. 7 (1956), 3-7.

Adolf Erik Nordenskiöld (1832-1901)

Swedish explorer and historian of cartography. Chief of the mineralogy division of the National Museum, Stockholm, 1858-1901.

George Kish in DSB, vol. 10 (1974), 148-149; Freeman, 318-319; James, 309n; George Kish, "Adolf Erik Nordenskiöld (1832-1901): Polar Explorer and Historian of Cartography," Geographical Journal, vol. 134 (1968), 487-500; George Kish, North-East Passage: Adolf Erik Nordenskiöld, His Life and Times (Amsterdam: Nico Israel, 1973).

Joseph Partsch (1851-1925)

German geographer. Professor of geography in the University of Leipzig, 1905-1922.

Dickinson, Makers, 89-93; H. Overbeck, "Joseph Partsch's Beitrag zur landeskundlichen Forschung," Berichte für Deutschen Landeskunde (January 1953), 34-56; obit. by F. W. Paul Lehmann in Geographische Zeitschrift, vol. 31 (1925), 321-329.

Siegfried Passarge (1866-1958)

German geographer. Professor of geography in the University of Breslau, 1905-1908, Colonial Institute in Hamburg, 1908-1919, and University of Hamburg from 1919.

Dickinson, Makers, 137-141; Fischer, 143-154; Freeman, 319; James, 234-235; Helmut Kanter, "Siegfried Passarges Gedanken zur Geographie," Die Erde, vol. 91 (1960), 41-51.

Albrecht Penck (1858-1945)

German geographer. Professor of geography in the University of Berlin, 1906-1926.

Dickinson, Makers, 100-111; Robert Beckinsale in DSB, vol. 10 (1974), 501-506; Fischer, 99-106; James, 233n; Herbert Louis, "Albrecht Penck und sein Einfluss auf Geographie und Eiszeitforschung," Die Erde, vol. 89 (1958), 161-182; Gerhard Engelmann, "Bibliographie Albrecht Penck," Wissenschaftliche Veröff. d. Deutschen Institut für Landeskunde (1960), 331-447; obit. by Johann Sölch in Mitteilungen der Geographischen Gesellschaft in Wien, vol. 89 (1946), 88-122.

Oscar Peschel (1826-1875)

German geographer. Professor of geography in the University of Leipzig, 1871-1875.

Friedrich Ratzel in ADB, vol. 25 (1887), 416-430; Dickinson, Makers, 55-59; James, 216-217.

August Petermann (1822-1878)

German geographer and cartographer. Associated with the Justus Perthes Geographical Institute, Gotha, 1854-1878.

H. Wichmann in ADB, vol. 26 (1888), 795-805; Freeman, 320; Werner Horn, "Die Geschichte der Gothaer Geographischen Anstalt im Spiegel des Schrifttums," Petermanns Geographische Mitteilungen, vol. 104 (1960), 271-287; Gerhard Engelmann, "August Petermann als kartographenlehrling bei Heinrich Berghaus in Potsdam," Petermanns Geographische Mitteilungen, vol. 106 (1962), 161-182.

Alfred Philippson (1864-1953)

German geographer. Professor of Geography in the University of Bonn, 1909-1929.

115

Dickinson, Makers, 142-144; Fischer, 107n; James, 238-239; E. Kirsten, "Alfredo Philippson (1865 [sic]-1953) e i suoi studi sui paesi mediterranei," Rivista Geografica Italiana, vol. 60 (1953), 467-470; Herbert Lehmann, "Alfred Philippson zum Gedächtnis anlässlich der 100. Wiederkehr seines Geburtstages am 1. Januar 1964," Geographische Zeitschrift, vol. 52 (1964), 1-6.

Philippe Pinchemel (b. 1923)

French geographer. Professor of geography in the University of Paris. Author of numerous works in urban geography and the geography of France. President of the Commission on the History of Geographical Thought of the International Geographical Union, 1968-1980. Editor of La Géographie à travers un siècle de congrès internationaux (1972) and co-editor of Geographers: Biobibliographical Studies, vols. 1-4 (1977-1980).

John Wesley Powell (1834-1902)

American geographer, geologist, and ethnologist. Head of the Bureau of American Ethnology, 1879-1902; Director of the U. S. Geological Survey, 1881-1894.

Wallace Stegner in DSB, vol. 11 (1975), 118-120; Freeman, 320; Preston James in Geographers, vol. 3 (1979), 117-124; James, 206-209; W. M. Davis, "Biographical Memoir of John Wesley Powell," National Academy of Science, Biographical Memoirs, vol. 8 (1915), 11-83; William Culp Darrah, Powell of the Colorado (Princeton: Princeton University Press, 1951).

Friedrich Ratzel (1844-1904)

German geographer. Professor of geography in the University of Leipzig, 1886-1904.

Crone, 34-35; Dickinson, Makers, 64-76; Robert Beckinsale in DSB, vol. 11 (1975), 308-310; Fischer, 95-99; Freeman, 320-321; Marvin Mikesell in International Encyclopedia of the Social Sciences, vol. 13 (1968), 327-329; Robert Lowie, The History of Ethnological Theory (New York, 1937), 119-127; Johannes Steinmetzler, "Die Anthropogeographie Friedrich Ratzels und ihre ideengeschichtlichen Wurzeln," Bonner geographische Abhandlungen, vol. 19 (1956), 1-151; Harriet Wanklyn, Friedrich Ratzel, A Biographical Memoir and Bibliography (Cambridge: Cambridge University Press, 1961); Carl O. Sauer, "The Formative Years of Ratzel in the United States," Annals of the Association of American Geographers, vol. 61 (1971), 245-254; Günther Buttmann, Friedrich Ratzel: Leben und Werk eines deutschen Geographen, 1844-1904 (Stuttgart: Wissenschaftliche Verlagsgesellschaft, 1977); obits. by Kurt Hassert in Geographische Zeitschrift, vol. 11 (1905), 305-325, 361-380, and by Jean Brunhes in La Géographie, vol. 10 (1904), 103-108.

Ernst Georg Ravenstein (1834-1913)

British geographer and cartographer. Topographical Department of the War Office, London, 1855-1872. Professor of geography in Bedford College, London, 1882-1883.

David Grigg in Geographers, vol. 1 (1977), 79-82; E. G. Ravenstein, A Life's Work (London: Privately printed, 1908); David Grigg, "The First English New Geographer," Geographical Magazine, vol. 46 (1974), 246-247; David Grigg, "E. G. Ravenstein and the 'Laws of Migration'," Journal of Historical Geography, vol. 3 (1977), 41-54; obit. by George Philip in Geographical Journal, vol. 41 (1913), 497-498.

Elisée Reclus (1830-1905)

French geographer, writer, and anarchist. Professor of geography in the New University of Brussels, 1894-1905.

Dickinson, Makers, 222-224; Gary Dunbar in DSB, vol. 11 (1975), 337-338; Fischer, 175-180; Freeman, 321; B. Giblin in Geographers, vol. 3 (1979), 125-132; James, 191-192; Gary Dunbar, Elisee Reclus, Historian of Nature (Hamden, Connecticut: Archon Books, 1978); obits. by Peter Kropotkin in Geographical Journal, vol. 26 (1905), 337-343, and by Patrick Geddes in Scottish Geographical Magazine, vol. 21 (1905), 490-496, 548-555.

Jacques Richard-Molard (1913-1951)

French geographer. Head of the Geography Section of the Institut Français d'Afrique Noire, Dakar, from 1945; Professor of geography in the Ecole coloniale (Ecole nationale de la France d'Outre-Mer), Paris, 1949-1951.

Hommage à Jacques Richard-Molard, 1913-1951 (Présence africaine, no. 15) (Paris, 1953) (2nd ed. revised appeared as Problèmes humains en Afrique occidentale, Paris, 1958); obits. by Théodore Monod in Bulletin de l'I.F.A.N., vol. 13 (1951), 953-957, and by Raoul Blanchard in Revue de géographie alpine, vol. 40 (1952), 7-16.

Ferdinand von Richthofen (1833-1905)

German geographer. Professor of geography in the University of Leipzig, 1883-1886, University of Berlin, 1886-1905.

Crone, 32-33; Dickinson, Makers, 77-88; Robert Beckinsale in DSB, vol. 11 (1975), 438-441; Fischer, 84-95; Freeman, 321; James, 217-220; Alfred Hettner, "Ferdinand von Richthofens Bedeutung für die Geographie," Geographische Zeitschrift, vol. 12 (1906), 1-11; Alfred Philippson, "Ferdinand von Richthofen als akademischer Lehrer," Geographische Zeitschrift, vol. 26 (1920), 257-272; Albrecht Penck, "Richthofens Bedeutung für die Geographie," Berliner geographische Arbeiten, no. 5 (1933), 1-17.

Carl Ritter (1779-1859)

German geographer. Professor of geography in the University of Berlin, 1820-1859.

Crone, 19-22; Dickinson, Makers, 34-48; Fischer, 65-78; Freeman, 321; Max Linke in Geographers, vol. 5 (1981), 99-108; James, 164-170; Carl Sauer in Encyclopedia of the Social Sciences, vol. 13 (1934), 395; Ernst Plewe in International Encyclopedia of the Social Sciences vol. 13 (1968), 517-520; Hanno Beck, "Carl-Ritter-Forschungen," Erdkunde, vol. 10 (1956), 227-233; Karl A. Sinnhuber, "Carl Ritter, 1779-1859," Scottish Geographical Magazine, vol. 75 (1959), 153-163; Fritz L. Kramer, "A Note on Carl Ritter, 1779-1859," Geographical Review, vol. 49 (1959), 406-409; Ritter bibliography in Die Erde, vol. 94 (1963), 13-36; Carl Ritter, "Introduction à la géographie générale comparée," ed. by Georges Nicolas-Obadia, Cahiers de géographie de Besançon, no. 22 (Paris, 1974); Hanno Beck, Carl Ritter, Genius der Geographie (Berlin: Dietrich Reimer Verlag, 1979); Manfred Büttner, editor, Carl Ritter: Zur europaisch-amerikanischen Geographie an der Wende vom 18. zum 19. Jahrhundert (Paderborn, etc.: Ferdinand Schöningh, 1980); obit. by Arnold Guyot in Journal of the American Geographical Society, vol. 2 (1860), 25-63.

Charles Robequain (1897-1963)

French geographer. Professor of colonial geography in the University of Paris from 1938.

Jean Delvert in "Géographes," 145-151; Meynier, 101.

Karl Theodor Sapper (1866-1945)

German geographer. Professor of geography in the universities of Tübingen (1902-1910), Strassburg (1910-1919), and Würzburg (1919-1932).

Dickinson, Makers, 145-147; Franz Termer, Karl Theodor Sapper, 1866-1945. Leben und Wirken eines deutschen Geographen und Geologen ("Lebensdarstellungen deutschen Naturforscher," no. 12) (Leipzig: Barth, 1966); obit. by Franz Termer in Petermanns Geographische Mitteilungen, vol. 92 (1948), 193-195.

Carl Ortwin Sauer (1889-1975)

American geographer. Professor of geography in the University of California, Berkeley, 1923-1957.

Dickinson, Reg. Con., 314-326; Fischer, 424-435; John Leighly in Geographers, vol. 2 (1978), 99-108; James, 399-402, 406; James Parsons in International Encyclopedia of the Social Sciences, vol. 14 (1968), 17-19; Gottfried Pfeifer, "Carl Ortwin Sauer zum 75. Geburtstage am 24.XII.1964," Geographische Zeitschrift, vol. 53 (1965), 1-9 (followed by anonymous article, "Die Berkeleyer Geographische Schule im Spiegel der unter Leitung und auf Anregung von C. O. Sauer hervorgegangen

118

Dissertationen 1927-1964," pp. 74-77). Obits. by Dan Stanislawski in _Journal of Geography_, vol. 74 (1975), 548-554, by Gottfried Pfeifer in _Geographische Zeitschrift_, vol. 63 (1975), 161-169, and by John Leighly in _Annals of the Association of American Geographers_, vol. 66 (1976), 337-348.

Gerhard Schott (1866-1961)

German oceanographer. Naval observatory, Hamburg, 1890-1931.

James, 237-238.

Ellen Churchill Semple (1863-1932)

American geographer. Special lecturer in geography, University of Chicago, 1906-1923; Professor of anthropogeography, Clark University, from 1921.

Dickinson, _Reg. Con._, 260-268 (from obituary by Colby, cited below); Fischer, 399-407; Freeman, 323; James, 377-381; Lawrence Gelfand, "Ellen Churchill Semple: Her Geographical Approach to American History," _Journal of Geography_, vol. 53 (1954), 30-41; John K. Wright, "Miss Semple's 'Influences of Geographic Environment': Notes toward a Bibliobiography," _Geographical Review_, vol. 52 (1962), 346-361; Mildred Berman, "Sex Discrimination and Geography: The Case of Ellen Churchill Semple," _Professional Geographer_, vol. 26 (1974), 8-11; Allen D. Bushong, "Women as Geographers: Some Thoughts on Ellen Churchill Semple," _Southeastern Geographer_, vol. 15 (1975), 102-109; obits. by Charles Colby in _Annals of the Association of American Geographers_, vol. 23 (1933), 229-240, and by R. H. Whitbeck in _Geographical Review_, vol. 22 (1932), 500-501.

Petr Petrovich Semenov-Tian-Shanskii (1827-1914)

Russian geographer, explorer, writer, and statistical compiler. Head of Central Statistical Committee from 1864; Senator, 1882; Member of State Council, 1897.

Vera Fedchina in _DSB_, vol. 12 (1975), 299-302; Iu. K. Efremov in _GSE_, vol. 23 (1979), 317-318; James, 284-285; David Hooson, "The Development of Geography in Pre-Soviet Russia," _Annals of the Association of American Geographers_, vol. 58 (1968), 258-259.

Jules Sion (1880-1940)

French geographer. Professor of geography in the University of Montpellier, 1910-1940.

Obituaries by Emm. de Martonne in _Annales de géographie_, vol. 49 (1940), 152-153, and by Lucien Febvre in _Annales d'histoire sociale_, vol. 3 (1941), 81-89.

Maximilien Sorre (1880-1962)

French geographer. Professor of geography in the University of Paris, 1941-1948.

Dickinson, <u>Makers</u>, 236-238; Fischer, 241-253; Pierre George in "Géographes," 185-195; Meynier, 52, 185; obit. by Pierre George in <u>Annales de géographie</u>, vol. 71 (1962), 449-459.

Laurence Dudley Stamp (1898-1966)

British geographer. Taught in the London School of Economics, 1926-1958 (Reader, 1926-1945, and Professor, 1945-1958).

Dickinson, <u>Reg. Con.</u>, 69-75 (from obituary by R. O. Buchanan in <u>IBG Stamp Memorial Volume</u>, 1970); James, 277-278; obit. by S. H. Beaver in <u>Geography</u>, vol. 51 (1966), 388-391.

Eduard Suess (1831-1914)

Austrian geologist. Professor of geology in the University of Vienna, 1867-1901.

E. Wegmann in <u>DSB</u>, vol. 13 (1976), 143-149; Freeman, 324; obits. by Norbert Krebs in <u>Mitteilungen der Geographischen Gesellschaft in Wien</u>, vol. 57 (1914), 296-311, and by P. Termier in <u>Revue générale des sciences pures et appliquées</u>, vol. 25 (1914), 546-552 (translated in <u>Smithsonian Annual Report for 1914</u>, 1915, 709-718).

Paul Vidal de la Blache (1845-1918)

French geographer. Professor of geography in the Ecole Normale Supérieure, 1877-1898, and the University of Paris, 1898-1914.

Crone, 23-26; Dickinson, <u>Makers</u>, 208-221; Fischer, 186-198; Freeman, 307-308; Philippe Pinchemel in "Géographes," 9-23 (reprint of article in <u>Geographisches Taschenbuch 1970/72</u>, Wiesbaden, 1972, 266-279); James, 246-253; Meynier, passim, espec. 17-35; Pierre Monbeig in <u>International Encyclopedia of the Social Sciences</u>, vol. 16 (1968), 316-318; T. W. Freeman, <u>The Geographer's Craft</u> (Manchester, 1967), chapter 3, "Vidal de la Blache, A Regional and Human Geographer," 44-71; Orlando Ribeiro, "En relisant Vidal de la Blache," <u>Annales de géographie</u>, vol. 77 (1968), 641-662; Paul Claval and Jean-Pierre Nardy, <u>Pour le cinquantenaire de la mort de Paul Vidal de la Blache: Etudes d'histoire de la géographie</u> (Cahiers de géographie de Besançon, 16) (Paris: Les Belles Lettres, 1968), see especially part 3, "Vidal de la Blache et la géographie française," by Paul Claval, 91-125; obit. by Lucien Gallois in <u>Annales de géographie</u>, vol. 27 (1918), 161-173.

Aleksandr Ivanovich Voeikov (1842-1916)

Russian geographer and climatologist. Professor in St. Petersburg University from 1885.

A. A. Fedoseev in DSB, vol. 14 (1976), 52-54; I. A. Fedosseyev in Geographers, vol. 2 (1978), 135-141; G. D. Rikhter in GSE, vol. 5 (1974), 546-547; James, 286-287; David Hooson, "The Development of Geography in Pre-Soviet Russia," Annals of the Association of American Geographers, vol. 58 (1968), 260-262; obits. by Wladimir Köppen in Petermanns Geographische Mitteilungen, vol. 62 (1916), 422-423, and Anonymous in Annales de géographie, vol. 25 (1916), 150-151.

Hermann Wagner (1840-1929)

German geographer. Professor of geography in the University of Göttingen, 1880-1920.

Dickinson, Makers, 93-94; Fischer, 113-114; Wagner's bibliography published in Petermanns Geographische Mitteilungen, vol. 66 (1920), 118-122, preceded by 3-page tribute to Wagner on the occasion of his 80th birthday by Paul Langhans; obits. by Ludwig Mecking in Geographische Zeitschrift, vol. 35 (1929), 585-596, and by W. Meinardus in Petermanns Geographische Mitteilungen, vol. 75 (1929), 225-229.

Moritz Wagner (1813-1887)

German geographer, zoologist, and naturalist-explorer. Professor and conservator of Ethnographic Museum in Munich.

Friedrich Ratzel in ADB, vol. 40 (1896), 532-543; Hanno Beck, "Moritz Wagner in der Geschichte der Geographie," Doctoral dissertation, University of Marburg, 1951; Hanno Beck, "Moritz Wagner als Geograph," Erdkunde, vol. 7 (1953), 125-127; Hanno Beck, chapter on Wagner in Grosse Reisende (Munich: Callwey, 1971), 190-208, 401-402.

Jacques Weulersse (1905-1946)

French geographer. Lecturer in geography in Paris at the Ecole nationale de la France d'Outre-Mer, Ecole nationale des langues orientales vivantes, Institut d'ethnologie, and Ecole libre des sciences politiques, 1941-1943; appointed professor of colonial geography in the University of Aix-Marseille, 1943.

Dickinson, Makers, 191, 234; Pierre Gourou in Geographers, vol. 1 (1977), 107-112; Meynier, 170-171; obit. by Elicio Colin (Weulersse's uncle) in Annales de géographie, vol. 56 (1947), 53-54.

John Kirtland Wright (1891-1969)

American geographer. Associated with the American Geographical
Society of New York, 1920-1956 (Librarian, 1920-1937, Director, 1938-
1949, Research Associate, 1949-1956).

Dickinson, Reg. Con., 304-313 (reprint of obituary by Bowden, below);
James, 416n; J. K. Wright, Human Nature in Geography: Fourteen Papers,
1925-1965 (Cambridge, Mass.: Harvard University Press, 1966), see
especially "Introduction," pp. 1-10; Wright bibliography published in
Geographies of the Mind: Essays in Historical Geosophy in Honor of
John Kirtland Wright, ed. by David Lowenthal and Martyn Bowden (New
York: Oxford University Press, 1976), 225-256; obits. by David
Lowenthal in Geographical Review, vol. 59 (1969), 598-604, and by
Martyn Bowden in Annals of the Association of American Geographers,
vol. 60 (1970), 394-403.